LET'S PREPARE FOR THE GRADE 8

MATH TEST

Anne M. Szczesny

Teacher and Administrator for the Gifted Math Program
University at Buffalo

BARRON'S

All inquiries should be addressed to:

Barron's Educational Series, Inc.
250 Wireless Boulevard
Hauppauge, New York 11788
http://www.barronseduc.com

International Standard Book No. 0-7641-1872-2

International Standard Serial No. 1534-326X

Printed in the United States of America
9 8 7 6 5 4

Contents

Chapter 1

The Test

Every student in New York State who entered Grade 8 in September 1998 or after is required to take the New York State Eighth Grade Math Test. Starting with the 2001 testing, seventh grade students may take the Eighth Grade Math Assessment Test if they are participating in an accelerated math or science program. Questions on the test are based on the math that you had up to and including eighth grade.

This test is used to see how well you

- listen and follow directions;
- understand major mathematical ideas;
- reason in real-world problems;
- use and apply formulas, definitions, signs, and symbols;
- read and interpret tables and graphs; and
- choose and use the appropriate procedure in solving a problem.

The math test is timed and will be given over two days. On the first day, the test has two parts. Part I has 27 multiple-choice questions (approximately 30 minutes) and Part II has four short and two longer open-ended response questions (approximately 30 minutes). On the second day, you will answer eight short and four longer open-ended response questions (approximately 60 minutes).

Each multiple-choice question will have four possible answers from which to choose the one correct answer. For open-ended response questions, you should provide an answer as well as show how you arrived at your answer. There is no *one* correct explanation for the open-ended response questions. So, be clear in your explanation. You will get partial credit on these problems.

Calculators are *not* allowed on the multiple-choice portion of the test. They are, however, to be used on the rest of the questions. You should use a scientific calculator. If you do not have one, use a four-function calculator ($+$, $-$, $\times$, $\div$) with a square-root key ($\sqrt{\ }$). A punch-out protractor and ruler as well as a reference sheet with some advanced formulas will be given to you when you need it.

SCORING

Each multiple-choice question is graded as either right or wrong. The open-ended response questions are graded using a rubric. The rubric is a set of guides the teachers follow to award points. The more you explain the process you used to get to your final answer, the more points you get. You can earn 0, 1, or 2 points on the shorter questions or 0, 1, 2, or 3 points on the longer ones.

Once your points have been tallied, your score for the test will be on a scale of 1 to 4, where 4 is the highest. A score of 4 or 3 shows that you exceed or meet the math standards, but a score of 2 or 1 shows that you need extra help or have serious deficiencies in your math knowledge and understanding. With a score of 1 or 2, you are required to take an extra class of math instruction. Your score will be recorded on your school records, but it will not be part of your class average. Your teacher will use your test results to see where you may need more study or practice in math.

TEST-TAKING TIPS

Here are a few test-taking tips to help you score well on the math test.

- Relax! You've taken timed tests before. This test gives you the opportunity to show your teachers how savvy you are in math!
- Listen carefully to and follow all the directions given by your teacher.
- Answer all the easy questions first. After you finish them, go back and work on the other questions.
- Make sure that you read each problem carefully and understand how you should answer the question. Gather all the information you need from the problem.
- Use your problem-solving techniques that you have learned, such as working backward, making lists, drawing pictures, guessing and testing, and breaking a large problem into smaller ones. When you get your answer, label it correctly with the proper measurements or units. For example, if you are answering a question that asks for the area of a triangle, label your answer with *square units* such as square inches or square meters.
- If you have extra time, go over your answers to make sure that they are correct. Write neatly because other teachers besides your math teacher will be scoring the tests.
- Do not leave any blanks! They will be scored as incorrect—zero points. Guess if you are not sure of an answer.

SAMPLE QUESTIONS

Practice questions will be provided at the end of each chapter. Here are samples of the types of questions you will be answering on the test.

MULTIPLE-CHOICE QUESTION

1. What is the prime factorization of 80?

    A. 2×40
    B. $5 \times 2 \times 4$
    C. $2^4 \times 5$
    D. $2 \times 5 \times 8$

Solution:
The correct answer is **C**. $80 = 2 \times 2 \times 2 \times 2 \times 5$ or $2^4 \times 5$. No partial credit is given to multiple-choice questions.

SHORT OPEN-ENDED RESPONSE QUESTION

2. Solve this equation for y. Show your work.

$$10 + 15 \div 5 \times 6 - 4 = y$$

Solution:
$$10 + (15 \div 5) \times 6 - 4 = y$$
$$10 + (3 \times 6) - 4 = y$$
$$10 + 18 - 4 = y$$
$$28 - 4 = y$$
$$24 = y$$

This algebraic solution is just one of many possible solutions. A point value of 2, 1, or 0 can be awarded for this type of question.

LONGER OPEN-ENDED RESPONSE QUESTION

3. To control her blood pressure, Jill's grandmother takes $\frac{1}{2}$ of a pill every other day. One supply of medicine contains 60 pills. How many months will one supply of pills last? Explain your answer.

   Solution:
   One-half of a pill every other day is another way of saying 1 pill every 4 days. At this rate, 60 pills will last 240 days. There are approximately 30 days in a month and 240 days is approximately 8 months. One supply of pills will last approximately 8 months.

   A point value of 3, 2, 1, or 0 can be awarded to answers of the longer open-ended response questions.

   This book was written to help you review some of those math topics you forgot or need to practice. Practice questions and solutions are at the end of each chapter. Remember that there is more than one way to solve a problem. You may have another solution or procedure that is mathematically sound but not presented in this book. In addition to the practice questions, a sample test is included with questions that are similar to the ones you may see on the test so you can time yourself.
   Remember that all New York State eighth grade students are taking the same test. Follow the examples and practice the Test Your Skills questions at the end of each chapter and the sample test. If there is something you don't understand, ask your teachers. They are there to help you. Good luck!

Chapter 2

Our Number System

NUMBER SETS

The first set people used to count their possessions was the *natural numbers,* or *counting numbers.*

$$\mathbb{N} = \{ 1, 2, 3, 4, 5, ...\}$$

To this set we added zero and formed the set of *whole numbers.*

$$\mathbb{W} = \{ 0, 1, 2, 3, 4, ...\}$$

Negative numbers, or *inverses,* were added to the whole numbers to form the set of *integers.* The letter $\mathbb{Z}$ represents this set. It is the first letter in the word *Zahlen,* the German word for number. Some math books may use $\mathbb{J}$ to represent the integers.

$$\mathbb{Z} = \{ ..., -3, -2, -1, 0, 1, 2, 3, ...\}$$

To represent parts of a whole, as in "half of a donut" or "three quarters of the distance," *rational numbers* come in to play. Rational numbers are not that easy to list. We use set builder notation to describe the elements of this set. This set includes fractions that can be written as decimals that terminate such as $\frac{1}{4}$ (or 0.25) and those whose digits repeat such as $\frac{2}{3}$ (or 0.666...).

$$\mathbb{Q} = \{ \tfrac{a}{b}, \text{ where } a, b \text{ are integers and } b \neq 0\}$$

The letter $\mathbb{Q}$ represents the set of rational numbers because it is the first letter in the word *quotient.* That's how the numbers are generated—through division.

The *irrational numbers* come next. These numbers can be written as decimals whose digits continue on without terminating or repeating a pattern, such as π (3.14159...) or $\sqrt{2}$ (1.414213...) or 5.105872898561234 ... . Remember that not all square roots are irrational numbers ($\sqrt{25} = \pm 5$).

A Venn diagram shows how all the sets fit together. The size of the circles representing the sets does *not* reflect the size of the sets, or the numbers of elements in each of the sets.

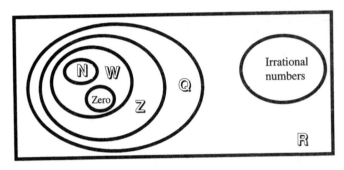

The ℝ in the bottom right corner of a Venn diagram shows the set you are working with. Here we are working with the set of real numbers. This set includes the rational numbers and the irrational numbers. We use real numbers in our everyday lives.

PROPERTIES OF NUMBERS

The operations of addition and multiplication are *commutative*. This means that, when we are adding or multiplying two numbers, the order of the numbers will not affect our answer.

$$a + b = b + a$$
$$3 + 7 = 7 + 3$$
$$10 = 10$$

$$c \times d = d \times c$$
$$6 \times 2 = 2 \times 6$$
$$12 = 12$$

The operations of addition and multiplication are also *associative*. When we add or multiply three numbers, we group two numbers with parentheses because these operations are binary (we work with two numbers at a time). No matter how we group these numbers, our answer will be the same. Notice that the three numbers remain in the same position. The parentheses do the grouping.

$$a + (b + c) = (a + b) + c$$
$$3 + (9 + 4) = (3 + 9) + 4$$
$$3 + 13 = 12 + 4$$
$$16 = 16$$

$$f \times (g \times h) = (f \times g) \times h$$
$$4 \times (5 \times 2) = (4 \times 5) \times 2$$
$$4 \times 10 = 20 \times 2$$
$$40 = 40$$

The operation of multiplication is *distributive over addition*. For example,

$$a \times (b + c) = (a \times b) + (a \times c)$$
$$2 \times (5 + 3) = (2 \times 5) + (2 \times 3)$$
$$2 \times 8 = 10 + 6$$
$$16 = 16$$

The *additive inverse* of the number 5 is −5 because when 5 is added to −5, the sum is zero.

$$a + (-a) = 0$$

The *multiplicative inverse* (or reciprocal) of a number multiplied by the number will always give you a product of 1. The reciprocal of $\frac{2}{5}$ is $\frac{5}{2}$, and vice versa, because their product is 1.

$$b \times \frac{1}{b} = 1$$

The *identity element for addition* is zero because any number added to zero results in the original number. If zero is added to 8, the sum is 8.

$$c + 0 = c$$

One is the *identity element for multiplication*. When muliplying a number by 1, your product is the original number. When 14 is multiplied by 1, the product is 14.

$$d \times 1 = d$$

The *Zero Product Property* states that any number multiplied by zero equals zero. When 24 is multiplied by 0, the product is 0.

$$g \times 0 = 0$$

OPERATIONS ON INTEGERS

ADDING INTEGERS

Positive + Positive = Positive   (7 + 4 = 11)
Negative + Negative = Negative   (−4 + −17 = −21)
Positive + Negative = Negative + Positive = Negative (Addition is commutative.)

Take the absolute value of the difference between the numbers. To find the absolute value of a number, look at the magnitude of the number while ignoring the sign. Then apply the sign of the number that has the greater absolute value.

Examples

   A. −16 + 4 = −12

16 − 4 = 12. Since |−16| > |4| and 16 is negative, so is the answer. Think of this problem as *owing* a friend $16. After you pay $4, you only *owe* $12. Start at −16 on the number line, move 4 spaces to the right, and arrive at your answer of −12.

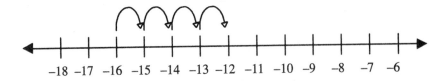

   B. −2 + 7 = 5

7 − 2 = 5. Since |7| > |−2| and 7 is positive, so is the answer.

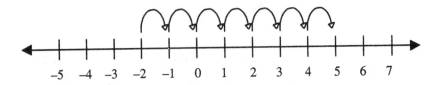

SUBTRACTING INTEGERS

When we subtract, we are really adding the inverse of the number that follows the operation sign. So, change the operation to addition, and change the second number to its inverse. Then, follow the rules for adding. If a number does not have a sign, it is a positive number. Here are some examples.

Examples

   A. 15 − 6 = 15 + −6 = 9

   B. −7 − 4 = −7 + −4 = −11

   C. 3 − 9 = 3 + −9 = −6

   D. 29 − (−7) = 29 + 7 = 36

   E. −12 − (−4) = −12 + 4 = −8

If you want to work on the number line, here is how subtracting works. Look at Example (E) above. Find –12 on the number line. The minus sign tells you *go to the left* when subtracting. But, the negative sign on 4 tells you go the opposite way, or else *go to the right*. So, you start going to the left (minus sign), but change directions (negative sign) and head to the right. You are actually adding a positive 4 to –12.

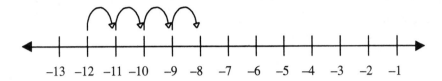

MULTIPLYING AND DIVIDING INTEGERS

Positive × Positive = Positive
Negative × Positive is the same as Positive × Negative = Negative
Negative × Negative = Positive

Examples

A. $7 \times 8 = 56$ (You have 7 boxes of crayons with 8 crayons in each.)

B. $-6 \times 8 = -48$ (I owe 6 cents to 8 friends.)

C. $6 \times -4 = -24$ (Over the last 6 months my locker rental bill was $4 per month. My total bill was $24.)

D. $-6 \times -3 = 18$ (If you went on a diet and *lost* 3 pounds per month, 6 months *ago* you were 18 pounds heavier.)

Since multiplication is the inverse operation of division, we know

$$\frac{positive}{positive} = positive \quad \left(\frac{72}{9} = 8\right) \qquad \frac{positive}{negative} = negative \quad \left(\frac{126}{-9} = -14\right)$$

$$\frac{negative}{negative} = positive \quad \left(\frac{-63}{-7} = 9\right) \qquad \frac{negative}{positive} = negative \quad \left(\frac{-54}{3} = -18\right)$$

Examples

A. $\dfrac{24}{3} = 8$ If you place 24 cupcakes on 3 plates, you have 8 per plate.

B. $\dfrac{15}{-5} = -3$ You go to the video arcade. You had \$15 five hours ago. Your loss rate is \$3 per hour.

C. $\dfrac{-35}{7} = -5$ Spending \$35 at an amusement park over 7 hours is the same as spending \$5 per hour.

D. $\dfrac{-42}{-6} = 7$ A person wants to lose 42 pounds and will try to lose 6 pounds per month. It should take her 7 months to lose the weight.

ORDER OF OPERATIONS

In the problem $15 + 9 \div 3 = x$, what is the value of x? The order we perform operations is

Parentheses, Exponents, Multiplication / Division, Addition / Subtraction

To help you remember this order, think of this mnemonic: **Please Excuse My Dear Aunt Sally.** The operations of multiplication and division are performed from left to right. The same goes for addition and subtraction. Perform the operations from left to right.

The correct answer to the problem is 18. First we divide; then we add. When everyone follows the same rules, everyone will get the same answer!

DIVISIBILITY RULES

Here are some ways to see if a number x is divisible by a number y without leaving a remainder. A number is divisible by

  2: if the number is even or the last digit is divisible by 2
  3: if the sum of the digits in the number is divisible by 3
  4: if the last two digits are divisible by 4
  5: if the last digit is 5 or 0
  6: if the number is even *and* the sum of the digits are divisible by 3
  8: if the last three digits are divisible by 8
  9: if the sum of the digits is divisible by 9
10: if the last digit is 0

PRIME FACTORIZATION AND EXPONENTS

When a factor is used more than once to obtain a product, we use a type of shorthand called exponents. For example $81 = 3 \times 3 \times 3 \times 3 = 3^4$. The 4 is called the *exponent* and the 3 is called the *base*. Three is used as a factor four times. If you don't see an exponent on a number, that factor is used once.

A *prime number* is a whole number greater than 1 and has exactly two different factors, itself and 1. The number 1 is *not* prime because it has only one factor, 1. We don't count the same factor twice. Two is the only prime number that is even. If a number has more than two factors, it is a *composite number*. Composite numbers can be written as a product of prime factors. We can break down composite numbers by creating a *factor tree*.

Examples

A.

B.

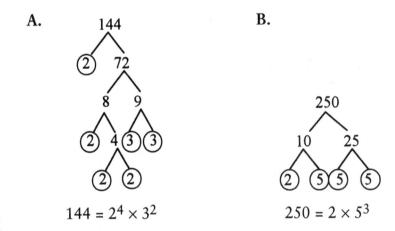

$$144 = 2^4 \times 3^2 \qquad\qquad 250 = 2 \times 5^3$$

SCIENTIFIC NOTATION

Very small or very large numbers can be written in scientific notation, another kind of shorthand. Numbers written in scientific notation have two parts: the first factor is a number between 1 and 10, and the second factor is a power of 10. For example 3,570,000,000 can be written as 3.57×10^9. The number 0.000000719 can be written as 7.19×10^{-7}.

The distance between the sun and Earth is approximately 93,000,000 miles or 9.3×10^7. A micrometer is one millionth of a meter. Most cells have a diameter of 5 to 12 micrometers, or, in scientific notation, 5×10^{-6} to 1.2×10^{-7} meters.

TEST YOUR SKILLS

1. Decide to which set the numbers below belong. A number may belong to more than one set. Explain your reasoning.

   A. $\sqrt{11}$

   B. $\dfrac{7}{28}$

   C. $\dfrac{-27}{3}$

   D. 18

   E. 7.302154387916...

2. Your friend does not understand the Distributive Property. Here is an example you made up to help her.

   Tamara has 5 bags, each containing 9 gem stones. She gives her sister 2 stones from each bag. Find the number of stones Tamara has now using the Distributive Property.

   Write an equation that shows how this property works.

3. True or False: The operation of addition is distributive over multiplication. Provide an example to back up your answer.

4. Given: $4 + 3(15 - 5) \div 6 + 2^3 = x$. Show all the steps you used in solving for x.

5. Reduce these fractions to lowest terms using the divisibility rules.

   A. $\dfrac{36}{243}$

   B. $\dfrac{32}{140}$

   C. $\dfrac{12}{60}$

6. Write the following numbers as a product of prime factors. Show your work by using a factor tree.
   A. 85
   B. 240
   C. 1323

7. Anne Boleyn was the second wife of Henry VIII. She is often referred to as Anne of a Thousand Days because that was how long she lived as Queen of England. Using scientific notation, represent the number of seconds she ruled as queen.

8. Which operation should you perform first in the following equation? Show all the steps in solving for x.

$$36 \div 9 + 5 \times 2 = x$$

9. Ms. Peters asked her students to substitute 3 for x in the expression $4x^2$. Dave said the expression was equal to 36, but Jill said it was equal to 144. Who is correct and why?

10. Sam opened up a checking account with $100.00. He intends to keep a minimum of $100.00 in his account at the end of every 6 weeks. Sam kept track of his weekly transactions.

Week #1	Withdrawal of $25.00 for Mom's birthday gift
Week #2	Deposit of $40.00 from paper route
Week #3	Deposit of $42.00 from grass-cutting business
Week #4	Withdrawal of $38.00 for bike replacement parts

   A. Was his balance at the end of the fourth week at least $100? Write an equation using integers for the transactions, then solve it.
   B. At this rate, how much will Sam have in his account at the end of the year?

TEST YOUR SKILLS SOLUTIONS

1. A. $\sqrt{11}$ is an irrational number because the digits in the decimal number repeat without a pattern. It is a real number because irrational numbers are real numbers.

   B. $\frac{7}{28} = \frac{1}{4}$ is a rational number because, when written as a decimal, one fourth is equal to 0.25. The digits terminate. It is a real number because rational numbers are real numbers.

   C. $\frac{-27}{3} = -9$ is an integer because it is a signed number. It is a rational number because you can write -9 as a ratio of two integers. It is a real number because integers are real numbers.

   D. 18 is a natural or counting number. It is a whole number because natural numbers are whole numbers. It is a rational number because you can write 18 as a ratio of two integers. It is a real number because natural numbers or whole numbers or integers are real numbers.

   E. 7.302143287916... is an irrational number because the digits do not terminate and there is no pattern to the digits. It is a real number because irrational numbers are real numbers.

2. $5 \times (9 - 2) = (5 \times 9) - (5 \times 2)$
   $$5 \times 7 = 45 - 10$$
   $$35 = 35$$

   Five represents the number of bags. Each bag contained nine stones before Tamara removed two stones from each bag to give to her sister.

3. $4 + (6 \times 3) = (4 + 6) \times (4 + 3)$
   $$4 + 18 = 10 \times 12$$
   $$22 \neq 120$$

   So, addition is *not* distributive over multiplication. This statement is false.

4.
$4 + 3(15 - 5) \div 6 + 2^3 = x$		Given equation
$4 + \quad 3(10) \div 6 + 2^3 = x$		Parentheses first $(15 - 5)$
$4 + \quad 3(10) \div 6 + \quad 8 = x$		Exponents next $(2^3 = 8)$
$4 + \quad\quad 30 \div 6 + \quad 8 = x$		Multiplication/Division from left to right (3×10)
$4 + \quad\quad\quad 5 \quad + \quad 8 = x$		Multiplication/Division from left to right $(30 \div 6)$
$9 + \quad 8 = x$		Addition/Subtraction from left to right $(4 + 5)$
$17 = x$		Addition/Subtraction from left to right $(9 + 8)$

5. **A.** $\frac{36}{243} = \frac{4}{27}$ Notice that the sum of the digits in both 36 and 243 is 9. Neither 4 nor 27 have common factors other than 1.

B. $\frac{32}{140} = \frac{8}{35}$ Both 32 and 140 are divisible by 4, but 8 and 35 have no common factors other than 1.

C. $\frac{12}{60} = \frac{4}{20} = \frac{1}{5}$ Both 12 and 60 are even, so divide by 2. We can divide by 4 because 4 and 20 have a common factor of 4.

6. **A.**

$$85 = 5 \times 17$$

B.

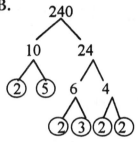

$$240 = 2^4 \times 3 \times 5$$

C.

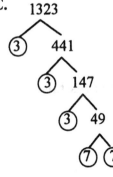

$$1323 = 3^3 \times 7^2$$

7. 60 seconds per minute × 60 minutes per hour = 3600 sec/hr

    3600 sec/hr × 24 hr/day = 86,400 sec/day

    86,400 sec/day × 1,000 days = 86,400,000 sec in 1000 days

    Queen Anne reigned for 8.64×10^7 seconds.

8. According to the order of operations, division is done first in the equation.

$$36 \div 9 + 5 \times 2 = x$$
$$4 + 5 \times 2 = x$$
$$4 + \quad 10 = x$$
$$14 = x$$

9. In this expression, 4 is multiplied by x^2. Since exponents are simplified before multiplication, $4x^2 = 4\,(3^2) = 4 \times 9 = 36$. So, Dave is correct.

10. **A.** Withdrawals can be written as negative numbers. Deposits are positive numbers. The equation is: $100 - 25 + 40 + 42 - 38 = 119$. So, Sam has $119 at the end of the month, including the original deposit of $100.

    B. An increase of $19 over 12 months is $228. When the original deposit is included, Sam can expect to have $328 in his account by the end of the year.

Chapter 3

Measurement

CUSTOMARY SYSTEM

Before there were instruments of measurement such as a ruler or meter stick, people used their body parts. For example, a yard was the distance between a person's nose and the end of his hand or a foot was the length of the measurer's foot. Since not all people are created "equal," a standardized system was developed. We in the United States use the customary system most of the time, but also use the metric system.

Customary System

Length	Weight	Capacity
12 inches = 1 foot	16 ounces = 1 pound	8 ounces = 1 cup
3 feet = 1 yard	2000 pounds = 1 ton	2 cups = 1 pint
5280 feet = 1 mile		2 pints = 1 quart
		4 quarts = 1 gallon

Examples

Using these basic units of length, answer these questions.

A. How many yards in a mile?

$$\frac{5,280\,^{ft}/_{mi}}{3\,^{ft}/_{yd}} = 1,760\,^{yd}/_{mi}$$ There are 1760 yards in a mile.

B. Can 6 cups of orange juice fit in a 1-quart container?

If 2 cups = 1 pint and 2 pints = 1 quart, then 4 cups = 1 quart. No, 6 cups will not fit in a 1-quart container.

C. Which is larger: $\frac{1}{4}$ of a ton or 600 pounds?

600 pounds is greater because $\frac{1}{4}$ of a ton is 500 pounds.

Examples

Using your ruler, answer these questions.

A. How thick is your math book to the nearest quarter inch?

B. What is the length of the pencil you are using now to the nearest inch?

C. How wide is your desk at school?

D. How long is your left sneaker?

E. What is the length of the line below?

_____

The first four questions will have different answers. The length of the line in the last question is 4 inches.

METRIC SYSTEM

The most widely used system of measurement in the world is the metric system. It is based on powers of ten. The prefixes help you remember the units of measure. The basic units of measure are *meter* (length), *gram* (mass), and *liter* (capacity).

Metric System

Unit	1/1000	1/100	1/10	1	× 10	× 100	× 1000
Length	**millimeter**	**centimeter**	decimeter	**meter**	dekameter	hectometer	**kilometer**
Mass	**milligram**	centigram	decigram	**gram**	dekagram	hectogram	**kilogram**
Capacity	**milliliter**	centiliter	deciliter	**liter**	dekaliter	hectoliter	kiloliter

The prefixes (milli — 1/1000, centi — 1/100, deci — 1/10, deka — 10, hecto — 100, and kilo — 1000) added to the basic units give you measures in the metric system. The most commonly used measures are in bold in the chart. A paperclip is about 1 gram. The doorknob is about 1 meter from the floor. A liter is a little larger than a quart. Because this system is based on powers of ten, it is easier to use than our customary system.

Examples

Answer these questions.

A. How many milliliters are in 80 liters?

Conversion is easy in this system. To get from milliliters to liters, you move three places to the right. You have 80.0 liters. So, move the decimal point three place to the right. There are 80,000 milliliters in 80 liters.

B. If one paperclip weights about 1 gram and the box about 3 grams, what is the weight in grams of six boxes of paperclips?

There are 6 boxes × 53 grams (50 grams for the paperclips + 3 grams for the box) or 318 grams.

C. How many meters are there in 735 centimeters?

To get from centimeters to meters, move two places to the left. In 735.0 cm there are 7.35 m.

Examples

Use a ruler or meter stick to answer these questions.

D. Find the height of your friend to the nearest tenth of a meter.

E. Approximate the distance between your elbow and wrist in centimeters.

F. Determine the number of pennies that can be laid in a straight row that measures 40 cm.

G. Find the length of your math book to the nearest millimeter.

H. Find the length of the following line.

The first four questions will have different answers. The line in the last question is 6.5 centimeters or 65 millimeters long.

CONVERTING BETWEEN SYSTEMS

Metric to Customary	Customary to Metric
$1 \text{ km} \approx \frac{5}{8} \text{ mi}$	$1 \text{ mi} \approx 1.6 \text{ km}$
$1 \text{ kg} \approx 2.2 \text{ lb}$	$1 \text{ lb} \approx 0.45 \text{ kg}$
$1 \text{ L} \approx 1.06 \text{ qt}$	$1 \text{ qt} \approx 0.95 \text{ L}$

Canadians use the metric system. Car speedometers show speeds in both systems. If you see a speed limit sign posted at 80 km/hr, you are traveling at approximately 50 mph. Prices at Canadian gas stations are shown as the cost per liter.

Suppose that the cost per Imperial gallon in Buffalo is $1.47 and the cost per Canadian liter is $0.43. Is the gas cheaper in Canada or in New York? One gallon of gas is approximately 4 times 0.95, or 3.8 liters. The cost per liter would be $0.39. Since the Canadian gas is $0.43 per liter, New York gas is cheaper.

TEST YOUR SKILLS

1. How many inches in one half of a yard?

2. If 1 gallon of water weights approximately 8 pounds, estimate the weight of 1 cup of water.

3. How many feet in one quarter of a mile?

4. How many 40-pound sacks of rice are there in three tons? Show your work.
   A. 120 sacks
   B. 150 sacks
   C. 60 sacks
   D. 80 sacks

5. Which metric measure would be best to use to list the weight of your bike? Explain your answer.

6. How many quarts of water are there in a 10-gallon aquarium? Show your work.

7. The room parents of the eighth grade class are planning a picnic. They have to decide on the amount of orange drink to bring. There are 24 students. Each student can have three servings. Each serving is 12 ounces. How many gallons should the parents bring? Show your work.

8. Monique has a birthday gift for her friend. She wants to wrap ribbon around a CD. Which amount would seem the best to use? Explain your answer.
   A. 200 mm
   B. 2 feet
   C. 6 inches
   D. 10 cm

9. Matt and Paul want to sign up for a 5-kilometer race. Because of his injured knee, Paul is not able to run distances longer than 3 miles. Should Paul sign up for the race? Explain your answer.

10. Megan's family is planning a camping vacation in Canada. They will fill up the gas tank in their van before they cross the border and won't fill up again until they get back to New York State. Megan's mom knows the range of their van is 440 miles on one tank of gas. The family wants to stay at a campground that is 270 km from their home. Will they make it back home without running out of gas? Explain your answer.

TEST YOUR SKILLS SOLUTIONS

1. There are 12 inches in 1 foot and 3 feet in one yard, so there are 36 inches in 1 yard. Half of 36 is 18. So there are 18 inches in one half of a yard.

2. One gallon contains four quarts, and each quart contains four cups. There are 16 cups in 1 gallon. If the gallon weights about 8 pounds, then 1 cup weights about $\frac{1}{2}$ pound or 8 ounces.

3. If there are 5280 feet in 1 mile, there are $\frac{5280}{4}$ or 1320 feet in a quarter of a mile.

4. One ton = 2000 pounds. There would be 2000 ÷ 40 or 50 sacks in 1 ton. Since there are 3 tons, there are 50 × 3 or 150 sacks. The answer is **B**.

5. You would weight your bike in pounds. Kilograms are the closest in measure to pounds.

6. There are 4 quarts in 1 gallon. This means that there are 40 quarts in 10 gallons.

7.
$$3 \text{ servings} \times 12 \text{ oz} = 36 \text{ oz per student}$$
$$36 \text{ oz} \times 24 \text{ students} = 864 \text{ oz}$$

There are 32 oz in 1 quart, which means there are 128 ounces in 1 gallon.

$$864 \text{ oz} \div 128 \text{ oz} = 6.75 \text{ gal}$$

If the parents bring 7 gallons of orange drink, there will be enough.

8. We know that 200 mm is about 8 inches and 10 cm is about 4 inches. The answer is **B**. Two feet is long enough to wrap around a CD whose case measures 5 by 5.5 inches. The other lengths will not be long enough to wrap around the case.

9. 1 km ≈ $\frac{5}{8}$ mi, so 5 km ≈ $\frac{25}{8}$ or $3\frac{1}{8}$ mi. This race is too long for Paul to run.

10. 440 mi × $\frac{8}{5}$ km/mi ≈ 704 km. The campground is 270 km one way or 540 km round trip. With a range of 704 km, there will be plenty of gas for their vacation.

Chapter 4

Fractions, Decimals, and Percents

A *fraction*, or rational number, represents the quotient between the number on top (numerator) and the number on the bottom (denominator). A decimal number, in which the digits can terminate or end (0.25) or repeat in a pattern (5.33333... or 0.284284284...) is another way to express a part of a number. Percents are based on *parts per hundred*.

Sometimes we need to solve problems that contain a combination of fractions, decimal numbers, and percents. The easiest thing to do is to change the numbers so that they are all in the same form.

CHANGING FRACTIONS TO DECIMALS

A fraction is a division problem. Divide the numerator by the denominator.

Examples

Change the following fractions to decimals.

A. $\frac{3}{5} = 3 \div 5 = 0.6$

B. $\frac{17}{20} = 17 \div 20 = 0.85$

C. $\frac{7}{9} = 0.7777... = 0.\overline{7}$

D. $\frac{2}{11} = 0.18181818... = 0.\overline{18}$

You may notice a pattern in the digits when dividing. Draw a line over the group of digits that repeat. Calculators will round or truncate. Suppose your calculator shows eight digits in the display. When dividing 7 by 9, your answer may look like 0.777777 if your calculator truncates (drops the rest of the digits), or 0.7777778 if your calculator rounds.

CHANGING FRACTIONS TO PERCENTS

When changing a fraction to a percent, change the given fraction to an equivalent fraction that has a denominator of 100. Write the numerator with a percent sign.

Examples

A. Change $\frac{4}{100}$ to a percent.

$$\frac{4}{100} = 4\%$$

B. Change $\frac{18}{25}$ to a percent.

$$\frac{18}{25} = \frac{72}{100} = 72\%$$

C. Change $\frac{3}{8}$ to a percent.

$$\frac{3}{8} = \frac{375}{1000} = \frac{37.5}{100} = 37.5\%$$

D. Change $\frac{2}{3}$ to a percent.

Sometimes it is easier to change the fraction to a decimal, then the decimal to a percent.

$$\frac{2}{3} = 2 \div 3 = 0.\overline{6} = 0.666... = 66.\overline{6}\%$$

CHANGING DECIMALS TO PERCENTS

To change decimals to percents, move the decimal point two places to the right and add a percent sign.

Examples

A. Change 0.47 to a percent.

$$0.47 = 47\%$$

B. Change 2.835 to a percent.

$$2.835 = 283.5\%$$

C. Change 0.009 to a percent.

$$0.009 = 0.9\%$$

D. Change 0.6 to a percent.

$$0.6 = 0.60 = 60\%$$

ORDERING FRACTIONS

When putting a group of fractions in order from least to greatest, first check the denominators.

If the *denominators are the same,* order the fractions by the numerator from least to greatest.

Example

A. Order from least to greatest: $\dfrac{11}{15}, \dfrac{2}{15}, \dfrac{7}{15}, \dfrac{13}{15}$.

$$\dfrac{2}{15}, \dfrac{7}{15}, \dfrac{11}{15}, \dfrac{13}{15}$$

If the *denominators are different,*

1. find the least common denominator (LCD) or the smallest multiple the denominators have in common;
2. change each fraction to its equivalent fraction using the LCD;
3. compare the numerators of the new fractions and order them; and then
4. write them in order using the original fractions.

Examples

B. Order from least to greatest: $\dfrac{1}{6}, \dfrac{3}{4}, \dfrac{11}{12}, \dfrac{2}{3}$.

The LCD of 6, 4, 12, and 3 is 12.

$$\dfrac{1}{6} \qquad \dfrac{3}{4} \qquad \dfrac{11}{12} \qquad \dfrac{2}{3}$$
$$\downarrow \qquad \downarrow \qquad \downarrow \qquad \downarrow$$
$$\dfrac{2}{12} \qquad \dfrac{9}{12} \qquad \dfrac{11}{12} \qquad \dfrac{8}{12}$$

Now order them according to the numerator:

$$\dfrac{2}{12}, \dfrac{8}{12}, \dfrac{9}{12}, \dfrac{11}{12}$$

The answer is

$$\dfrac{1}{6}, \dfrac{2}{3}, \dfrac{3}{4}, \dfrac{11}{12}$$

C. Order from least to greatest: $\dfrac{5}{7}$, $\dfrac{1}{2}$, $\dfrac{6}{5}$, $\dfrac{1}{10}$.

Did you notice that $\dfrac{6}{5}$ is greater than one and the other three fractions are less than one? So, $\dfrac{6}{5}$ is the greatest and goes last. Now work with the other three. The LCD of 7, 2, and 12 is 70.

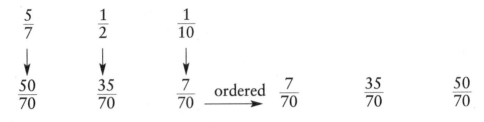

$$\dfrac{5}{7} \qquad \dfrac{1}{2} \qquad \dfrac{1}{10}$$

$$\downarrow \qquad\quad \downarrow \qquad\quad \downarrow$$

$$\dfrac{50}{70} \qquad \dfrac{35}{70} \qquad \dfrac{7}{70} \xrightarrow{\text{ordered}} \dfrac{7}{70} \qquad \dfrac{35}{70} \qquad \dfrac{50}{70}$$

The answer is

$$\dfrac{1}{10}, \quad \dfrac{1}{2}, \quad \dfrac{5}{7}, \quad \dfrac{6}{5}$$

D. Order from least to greatest: $\dfrac{7}{10}$, $\dfrac{1}{5}$, $\dfrac{3}{8}$, $\dfrac{2}{3}$.

You can change the fractions to decimals and then compare and order them.

$$\dfrac{7}{10} \qquad \dfrac{1}{5} \qquad \dfrac{3}{8} \qquad \dfrac{2}{3}$$

$$\downarrow \qquad\quad \downarrow \qquad\quad \downarrow \qquad\quad \downarrow$$

$$0.7 \qquad 0.2 \qquad 0.375 \qquad 0.\overline{6} \xrightarrow{\text{ordered}} 0.2,\ 0.375,\ 0.\overline{6},\ 0.7$$

So the answer is

$$\dfrac{1}{5}, \quad \dfrac{3}{8}, \quad \dfrac{2}{3}, \quad \dfrac{7}{10}$$

RATIO AND PROPORTION

A *fraction* is a *ratio*. Two ratios connected with an equal sign form a *proportion*. Proportions are used to solve many problems including those involving scale and measurement. If you know three of the four numbers, you can easily solve the proportion.
  Here are two identical circles.

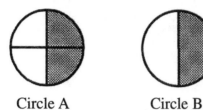

Circle A   Circle B

No one can argue that the same amount is shaded in both circles—one half. We can say that $\frac{2}{4}$ of circle A is shaded and $\frac{1}{2}$ of circle B is shaded.

$$\frac{2}{4} \bowtie \frac{1}{2}$$

The cross products are equal: $2 \times 2 = 4 \times 1$. Use this technique to check if the proportion is accurate.

Examples

A. The architect's plans are in the scale of $\frac{1}{3}$, which means 1 inch on the plans equals 3 feet in real life. How wide is the garage if the plans measure 6 inches?

$$\frac{1}{3} = \frac{6}{x}$$

In this proportion, both numerators represent the scale from the plans, while both denominators give the actual size. We can solve this proportion using cross products.

$$1 \times x = 6 \times 3$$
$$x = 18$$

So, if 1 inch represents 3 feet, then 6 inches represent 18 feet, and the garage is 18 feet wide.

B. The ship *Titanic* was 882 feet long. A scale model of the *Titanic* is $\frac{1}{63}$ of the actual size. Represent a proportion to calculate *l*, the length of the model in feet.

$$\frac{\text{``Inch'' scale of the model}}{\text{``Feet'' scale of the real ship}} = \frac{\text{Total ``inch'' length of the model}}{\text{Total ``feet'' length of the real ship}}$$

See how the proportion refers to the "model" in the numerator and the "real ship" in the denominator of *both* fractions. Using this proportion, we can set up the following proportion:

$$\frac{1}{63} = \frac{l}{882}$$

SOLVING PERCENT PROBLEMS

We can use our technique of solving proportions when working on percent problems. We know that $\frac{is}{of} = \frac{rate}{100}$. Read the problem carefully, then substitute the numbers into the proportion. Solve for the missing number.

Examples

A. What number is 35% of 140?

$$\frac{x}{140} = \frac{35}{100}$$

Cross multiply:

$$100x = 140 \times 35$$
$$100x = 4900$$
$$x = 49$$

Check your answer: does 35% of 140 equal 49?

$$0.35 \times 140 = 49$$
$$\therefore 49 \text{ is } 35\% \text{ of } 140$$

B. Fifty-four is what percent of 90?

$$\frac{54}{90} = \frac{x}{100}$$

Cross multiply.

$$90x = 54 \times 100$$
$$90x = 5400$$
$$x = 60$$

Check your answer: 60% of 90 should be 54.

$$0.60 \times 90 = 54$$
$$\therefore 60\% \text{ of } 90 \text{ is } 54$$

C. 48 is 12% of what number?

$$\frac{48}{x} = \frac{12}{100}$$

Cross multiply.

$$\frac{48}{x} = \frac{12}{100}$$

$$12x = 48 \times 100$$
$$12x = 4800$$
$$x = 400$$

Check your answer: 48 should be 12% of 400.

$$0.12 \times 400 = 48$$
$$\therefore 12\% \text{ of } 400 \text{ is } 48$$

ADDING AND SUBTRACTING FRACTIONS

When the denominators are the same, you can add or subtract fractions by adding or subtracting the numerators. Be sure to keep the same denominator. Reduce the fraction to lowest terms if needed.

Examples

A. $\dfrac{5}{12} + \dfrac{1}{12} = \dfrac{6}{12} = \dfrac{1}{2}$

B. $\dfrac{19}{27} - \dfrac{16}{27} = \dfrac{3}{27} = \dfrac{1}{9}$

C.
$$5\frac{3}{7}$$
$$+ \ 6\frac{6}{7}$$
$$\overline{11\frac{9}{7}} = 12\frac{2}{7}$$

D.
$$2\frac{3}{8}$$
$$- \ 1\frac{1}{8}$$
$$\overline{1\frac{2}{8}} = 1\frac{1}{4}$$

E.
$$7\frac{3}{11} = 6\frac{14}{11}$$
$$- \ 3\frac{6}{11} = 3\frac{6}{11}$$
$$\overline{\qquad\qquad 3\frac{8}{11}}$$

You cannot subtract $\frac{6}{11}$ from $\frac{3}{11}$, so rename $7\frac{3}{11}$ as $6\frac{14}{11}$ and subtract as usual.

When fractions have different denominators, find the LCD, and then rewrite equivalent fractions using the LCD.

Examples

F. $\quad \frac{5}{12} + \frac{1}{4} =$

$\quad \frac{5}{12} + \frac{3}{12} = \frac{8}{12} = \frac{2}{3}$

G. $\quad \frac{3}{4} + \frac{1}{5} + \frac{1}{2} =$

$\quad \frac{15}{20} + \frac{4}{20} + \frac{10}{20} = \frac{29}{20} = 1\frac{9}{20}$

H. $\quad \frac{1}{2} - \frac{3}{7} =$

$\quad \frac{7}{14} - \frac{6}{14} = \frac{1}{14}$

I. $\quad\quad 2\frac{1}{8} \quad = \quad 2\frac{3}{24}$

$\quad + \ 1\frac{2}{3} \quad = \quad 1\frac{16}{24}$

$\quad\quad\quad\quad\quad\quad\quad\quad 3\frac{19}{24}$

J. $\quad\quad 5\frac{1}{7} \quad = \quad 5\frac{6}{42} \quad = \quad 4\frac{48}{42}$

$\quad - \ 2\frac{5}{6} \quad = \quad 2\frac{35}{42} \quad = \quad 2\frac{35}{42}$

$\quad\quad\quad\quad\quad\quad\quad\quad\quad\quad\quad\quad\quad 2\frac{13}{42}$

MULTIPLYING AND DIVIDING FRACTIONS

The product of fractions is obtained by multiplying the numerators together to get a new numerator and then multiplying the denominators together to get a new denominator. Change mixed numbers to improper fractions; then multiply.

Examples

A. $\dfrac{2}{9} \times \dfrac{3}{4} \times \dfrac{2}{3} = \dfrac{12}{108} = \dfrac{1}{9}$

B. $3\dfrac{5}{7} \times 5\dfrac{1}{8} =$

$\dfrac{26^{13}}{7} \times \dfrac{41}{\not{8}_{4}} = \dfrac{533}{28} = 19\dfrac{1}{28}$

When dividing fractions, simplify your problem by changing the divisor (the number after the division sign) into its *reciprocal*. For example, the reciprocal of $\frac{2}{5}$ is $\frac{5}{2}$ because their product is 1. Change the division sign into a multiplication sign. Now you have a multiplication problem.

Examples

C. Find the reciprocals of the following numbers:

$\dfrac{5}{11}$

The reciprocal is $\frac{11}{5}$.

$\dfrac{14}{9}$

The reciprocal is $\frac{9}{14}$.

$2\dfrac{3}{5} = \dfrac{13}{5}$

The reciprocal is $\frac{5}{13}$.

8

The reciprocal is $\frac{1}{8}$.

D. $\dfrac{3}{7} \div \dfrac{6}{11} =$

$\dfrac{\cancel{3}^{1}}{7} \times \dfrac{11}{\cancel{6}_{2}} = \dfrac{11}{14}$

Leave the dividend (the first number) alone. Change the sign to multiplication. Change the divisor (the number *after* the sign) into its reciprocal. Reduce if you can. Multiply as usual.

E. $3\dfrac{1}{5} \div 4 =$

$\dfrac{\cancel{16}^{4}}{5} \times \dfrac{1}{\cancel{4}_{1}} = \dfrac{4}{5}$

Change the mixed number to an improper fraction. Then multiply by the reciprocal of the divisor.

TEST YOUR SKILLS

1. Shade in the remaining area so that a total of 80% is shaded. Explain your answer.

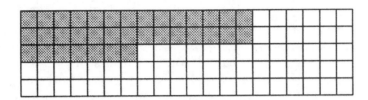

2. At the pizza party, Shelly ate $\frac{3}{10}$ of the pizza, and Kelly ate 20%. How much did they eat all together? Show your work.

3. Three students worked on a computer project. Sarah completed $\frac{2}{5}$ of the work and Tim completed 35% of the work. What fraction of the work was left for Carl to finish?

4. Solve for t: $\frac{4}{t} = \frac{2}{9}$

5. The directions on the container of iced tea mix state that you should use 1 cup of mix to make 1 quart of iced tea. One quart is equal to 32 ounces. The only container you have will hold 28 ounces. How much mix should you use to make 28 ounces of iced tea? Explain your answer.

6. 81 is what percent of 150?

7. Which choice listed below has the fractions in the correct order from least to greatest?

    A. $\frac{3}{8}, \frac{1}{4}, \frac{4}{7}, \frac{3}{5}$

    B. $\frac{3}{5}, \frac{1}{4}, \frac{4}{7}, \frac{3}{8}$

    C. $\frac{1}{4}, \frac{3}{8}, \frac{4}{7}, \frac{3}{5}$

    D. $\frac{1}{4}, \frac{3}{5}, \frac{4}{7}, \frac{3}{8}$

8. The Smithfield Middle School has an enrollment of 700 students. One day during the flu season, 95 students were absent. The school can close due to illness if 15% of the students are absent. Could Smithfield close that day? Explain your answer.

9. Using squares, draw a representation of $\frac{17}{4}$.

10. Howie went shopping for a new calculator. The first store's price was $35.95. There was a 10% off sale. Sales tax was 8%. The second store had the same calculator for $39.95 at 20% and a "No Sales Tax" special. Which store had the better offer? Explain your answer.

TEST YOUR SKILLS SOLUTIONS

1. $80\% = \dfrac{80}{100} = \dfrac{8}{10} = \dfrac{4}{5}$

There are 100 small squares in the rectangle. A total of 80 squares should be shaded. Thirty-five are done for you. You should shade in another 45 squares.

Here's another way to solve this problem. The rectangle shown here is split into five sections, each containing 20 squares. Four of the five sections must be shaded to show 80%.

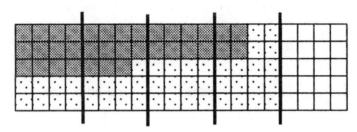

2. Shelly ate $\dfrac{3}{10}$ of the pizza, and Kelly ate 20%. Change both numbers to fractions or both to percents.

Solving with fractions:
$$20\% = \dfrac{20}{100} = \dfrac{2}{10} = \dfrac{1}{5}$$

Add together the girls' fractions:
$$\dfrac{3}{10} + \dfrac{2}{10} = \dfrac{5}{10} = \dfrac{1}{2}$$

Together they ate half of the pizza.

Solving with percents:
$$\dfrac{3}{10} = \dfrac{30}{100} = 30\%$$

Add 20% and 30% to get 50% of the pizza. Therefore, the girls ate 50% or $\dfrac{1}{2}$ of the pizza. Since this problem did not specify to give your answer as a percent or fraction, either answer is acceptable.

3. Sarah's work:

$$\frac{2}{5} = \frac{8}{20}$$

Tim's work:

$$35\% \text{ or } \frac{35}{100} = \frac{7}{20}$$

Sarah + Tim's work:

$$\frac{8}{20} + \frac{7}{20} = \frac{15}{20}$$

The whole project:

$$1 = \frac{20}{20}$$

Carl's work = The whole project − (the work done by Sarah + Tim)

$$\frac{20}{20} - \frac{15}{20} = \frac{5}{20} = \frac{1}{4}$$

Therefore, Carl had $\frac{1}{4}$ of the project to finish. (This problem asked that the answer be a fraction.)

4. $\frac{4}{t} = \frac{2}{9}$

Use cross multiplication.

$$4 \times 9 = 2t$$
$$36 = 2t$$

Divide both sides of the equation by 2.

$$t = 18$$

5. You need 1 cup of mix to make 1 quart of iced tea. There are 4 cups in 1 quart and 8 ounces in one cup. Set up a proportion.

$$\frac{8 \text{ oz of mix}}{32 \text{ oz of mix}} = \frac{x \text{ of drink}}{28 \text{ of drink}} \iff \frac{1}{4} = \frac{x}{28}$$

Solving for x, you get 7. Therefore, 7 oz of mix will make 28 oz of iced tea.

6. You are looking for the percent or rate in this problem. Use $\frac{is}{of} = \frac{rate}{100}$ and then substitute numbers.

$$\frac{81}{150} = \frac{r}{100}$$

Cross multiply.

$$81 \times 100 = 150r$$
$$8100 = 150r \quad \text{(Divide both sides by 150)}$$
$$54 = r$$

Therefore, 54% of 150 is 81.

7. Change the fractions into decimal form:

$$\frac{3}{8}, \quad \frac{1}{4}, \quad \frac{4}{7}, \quad \frac{3}{5}$$

$$\downarrow \qquad \downarrow \qquad \downarrow \qquad \downarrow$$

$$0.375 \quad 0.25 \quad 0.571... \quad 0.6 \xrightarrow{\text{ordered}} 0.25, \ 0.375, \ 0.571..., \ 0.6$$

Therefore, choice **C** is correct.

8. You need to find the number of students who must be absent before the school can close. That number is 15% of 700. When you see the word *of*, use multiplication.

$$0.15 \times 700 = 105$$

No, the school cannot close because 105 students needed to be absent for the school to close, and only 95 students are absent.

9. Rewrite the improper fraction as a mixed number.

$$\frac{17}{4} = 4\frac{1}{4}$$

Draw five squares. Shade in four whole squares and one quarter of the fifth square.

10. The first store:

	Cost is	$35.95
	10% off	− 3.60
		$32.35
	8% tax	+ 2.59
	Total:	$34.94

The second store:

	Cost is	$39.95
	20% off	− 7.99
	Total:	$31.96

Therefore, the second store has the better deal.

Chapter 5

Geometry

LINES

A *point* is a location in space. It has neither height nor length. If there are at least two points on the plane, there is a *line*. A line is a set of points that follow a path, which continues indefinitely in both directions. A portion of a line that has two endpoints is a *line segment*. A *ray* is part of a line that has only one endpoint. In this section, we will deal with straight lines, segments, and rays. Lines are identified by any two points on the line. The points are written as uppercase letters, but a line can also be given a name using a single lowercase letter. A line segment is identified by its two endpoints, and a ray is labeled with the endpoint, which is written first, and another point on the ray.

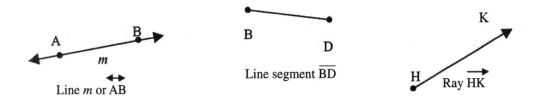

Line *m* or AB   Line segment BD   Ray HK

Lines are *parallel* if they are equidistant (the same distance apart) from each other. Right angles are formed when two lines intersect and are *perpendicular* to each other. The symbol for a right angle is the little square near the vertex of the angle (endpoint shared by two segments that form the angle).

Parallel lines   Perpendicular lines

ANGLES

When segments, rays, or lines meet or intersect, angles are created. Angles whose measure is between 0° and 90° are called *acute angles*. A 90° angle is called a *right angle*. If the measure of an angle is between 90° and 180°, the angle is an *obtuse angle*. An angle whose measure is exactly 180° is called a *straight angle*.

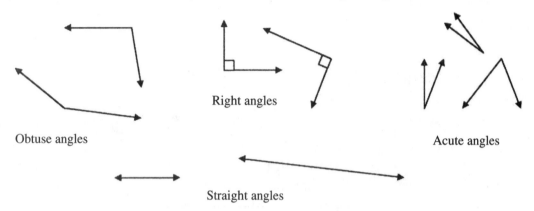

Obtuse angles   Right angles   Acute angles

Straight angles

A *protractor* is used to measure angles. There are two rows of numbers on a protractor. Place the crosshairs of the protractor on the vertex of the angle so that the horizontal line of the crosshairs is on one of the rays of the angle. The other ray of the angle will fall on one of the numbers of the protractor. For example, suppose the ray lands on the numbers 70° and 110°. The measure of the angle is 70° if your angle is acute or 110° if you are measuring an obtuse angle.

Examples

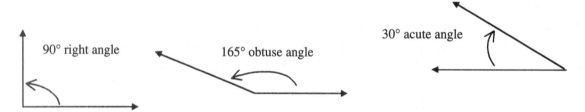

90° right angle   165° obtuse angle   30° acute angle

Use your protractor to verify these measurements. The rays that form angles continue infinitely. Extend the rays if you need more length for measuring an angle.

If the sum of two angles is 90°, the angles are *complementary angles*. Think of <u>com</u>-plementary and <u>c</u>orner. The angles form a right angle like the inside corner of a room. The complement of a 67° angle is a 23° angle (67° + 23° = 90°). The complement of a 12° angle is a 78° angle (12° + 78° = 90°).

If the sum of two angles is 180°, the angles are *supplementary* angles. Think of <u>supple</u>-mentary and <u>s</u>traight. The angles form a straight angle. The supplement of a 146° angle is a 34° angle (146° + 34° = 180°). The supplement of an 89° angle is a 91° angle (89° + 91° = 180°).

Two intersecting lines form four angles. *Vertical angles* are congruent in measure. Angles *a* and *b* are vertical angles as are angles *c* and *d*.

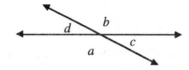

When two parallel lines are cut by a transversal (the line that cuts across the parallel lines), eight angles are formed.

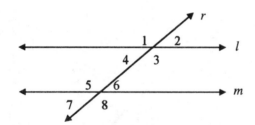

Angles 4 and 6, and angles 3 and 5 are called *alternate interior angles*. These angles are congruent in measure. You can see why if you imagine moving the bottom line so that it lies on the top line. Angle 6 lands on angle 2, and you know angles 2 and 4 are congruent because they are vertical angles. If line *l* is placed over line *m* the angles formed with the transversal will lay on top of each other. These *corresponding angles* are 1 and 5, 2 and 6, 3 and 8, and 4 and 7.

TWO-DIMENSIONAL FIGURES

TRIANGLE

All triangles have three angles and three sides that we use to classify them. An *equilateral triangle* has three equal sides with all three angles equal in measure. An *isosceles triangle* has two sides of the same length. The two base angles will have the same measure. A *scalene triangle*'s sides are all different in length.

   When combining the terms *acute, obtuse,* and *right* with *equilateral, isosceles,* and *scalene,* you can describe any triangle.

Examples

Identify these triangles using these terms.

A.

   This is an equilateral triangle because all three sides are equal in length, and the three angles are equal in measure. All equilateral triangles are acute.

B.

   This is a right isosceles triangle because of the right angle and two sides are equal in length.

C.

   This is an obtuse scalene triangle because there is an obtuse angle and no sides are equal in length.

D.

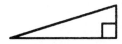

This is a right scalene triangle because there is a right angle and no sides are equal in length.

E.

This is an acute isosceles triangle because the largest angle is acute and two side are equal in length.

PARALLELOGRAMS

If a figure has four sides and the two pairs of opposite sides are parallel, then the figure is a *parallelogram*. *Rectangles* have two pairs of parallel sides and four right angles. *Squares* are special rectangles that have four equal side lengths. A *rhombus* looks like a square that was squeezed so that its shape resembles a diamond on a playing card.

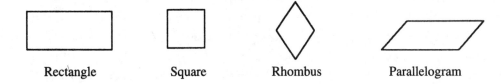

Rectangle   Square   Rhombus   Parallelogram

TRAPEZOID

A *trapezoid* has four sides but is not a parallelogram because it only has one pair of parallel sides. Depending on how it is constructed, a trapezoid can have a right angle or it can be an isosceles trapezoid where the nonparallel sides are equal in length and the base angles are equal.

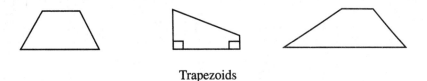

Trapezoids

OTHER POLYGONS

Polygon means *many angles*. *Poly-* is Greek for many and *-gon* comes from the Greek work for knee or bend or angle. Polygons are figures with many angles. They get their names from the number of angles they have. However, since they have the same number of angles as sides, we are more familiar with identifying them by their number of sides.

Number of Angles	Name of Figure
3	Triangle
4	Quadrilateral
5	Pentagon
6	Hexagon
7	Septagon
8	Octagon
9	Nonagon
10	Decagon

The list of polygon names goes on. Do your own research to find their names.

If the lengths of the sides of a certain polygon are the same, that figure is called a *regular polygon*. Regular triangles and rectangles have special names. A regular triangle is called an *equilateral triangle*. This word contains the Latin prefix *equi-* or equal. A regular rectangle is called a square. All other figures have the word *regular* in front of the name, such as regular hexagon and regular octagon, when the lengths of their sides are equal.

CIRCLE

If you marked a point (call this point A) on a sheet of paper and from A placed a point 4 cm away and continued to place points 4 cm from point A in all directions, you would get a circle. Circles are identified by their center point, in this case circle A. Each circle has radii (plural for radius), diameters, chords, and arcs. A *radius* is the distance between the center point and any other point on the circle. A *chord* connects any two points on the circle. The *diameter* is a special chord that cuts the circle in two halves *and* passes through the center point. An *arc* is a portion of the circle's circumference. There are 360° in a circle.

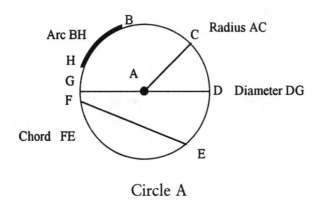

Circle A

Example

List the following parts of circle A.

    A. Radii   AC, AD, AG

    B. Diameter   DG

    C. Arcs   BC, CD, DE, EF, FG, GH, BD, BE, BF, BG, BH, CE, CF, CG, CH, DF, DG, DH, EG, EH, FH

    D. Chords   FE, DG

THREE-DIMENSIONAL FIGURES

RECTANGULAR SOLIDS

Wood that is used to build a house is called dimensional lumber. It has three measurements: height, width, and length. For example, inside your bedroom walls might be two-by-fours or two-by-sixes, with the length of this lumber cut from floor to ceiling.

Another example of a rectangular solid is a tissue box. It is composed of six faces, or three pairs of identical rectangles. A special rectangular solid that has 6 identical faces and 12 edges of equal length is called a cube.

Other nonrectangular, three-dimensional figures are spheres, pyramids, prisms, cylinders, and cones.

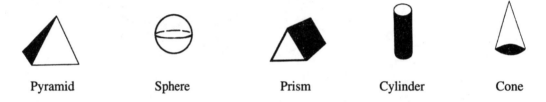

| Pyramid | Sphere | Prism | Cylinder | Cone |

SIMILARITY AND CONGRUENCE

Two triangles are similar if their corresponding angles are equal and pairs of corresponding sides have the same ratio. Triangle ABC has sides with lengths of 3, 4, and 5 units.

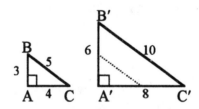

Triangle A′B′C′ has legs with lengths of 6, 8, and 10 units. Angles A and A′ are 90°, $\angle B \cong \angle B'$, and $\angle C \cong \angle C'$. AB and A′B′ are corresponding sides. Their ratio is one half. The ratio of the other pairs of corresponding sides will be the same.

$$\frac{AB}{A'B'} = \frac{AC}{A'C'} = \frac{BC}{B'C'} \text{ and } \frac{3}{6} = \frac{4}{8} = \frac{5}{10}$$

Examples

A. Which of these figures are similar?

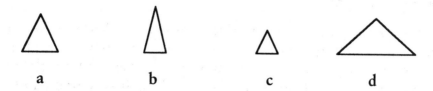

The triangles in **a** and **c** are similar.

B. Find the lengths of the missing sides of △KLM.

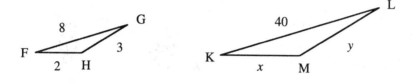

$$\frac{FG}{KL} = \frac{8}{40} = \frac{1}{5}$$

So, the ratio is $\frac{1}{5}$.

$$\frac{FH}{KM} = \frac{2}{x} = \frac{1}{5}$$

So, KM is 10 units.

$$\frac{GH}{LM} = \frac{3}{y} = \frac{1}{5}$$

So, LM is 15 units.

Congruent figures are exact copies of each other. Congruence applies not only to closed figures but to line segments and angles. For example, we can say that all right angles are congruent because the measure of the angles will always be 90°.

INTERIOR ANGLES OF POLYGONS

The following table shows the sum of the interior angles of geometric figures.

Name of Figure	Number of Sides	Number of Angles	Sum
Triangle	3	3	180°
Quadrilateral	4	4	360°
Pentagon	5	5	540°
Hexagon	6	6	720°
Septagon	7	7	900°
Octagon	8	8	1080°
n-gon	n	n	$(n-2)\,180°$

Example

C. In this isosceles triangle, $\angle QPR$ measures 66°. Find the measure of $\angle PQR$.

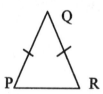

Since $\overline{PQ} = \overline{QR}$, then $\angle QPR \cong \angle QRP$. The sum of the interior angles of a triangle is 180°. We know the measure of two of the angles. Now we find the third.

$$2\,(66°) + x = 180°$$
$$132° + x = 180°$$
$$x = 48°$$

So, $\angle PQR = 48°$.

BASIC GEOMETRIC CONSTRUCTIONS

Geometry has been studied for thousands of years by different civilizations. The Greek mathematician Euclid is famous for his 13-volume set of books called *The Elements*. At the time it was written, *The Elements* contained all the known mathematics in the world. Much of what we know about geometry today is contained in *The Elements*.

The Greeks used a compass and straightedge for constructions. They did not have the accurate instruments we use today.

Examples

A. Find the perpendicular bisector of the line segment AB.

Use the endpoints of the segment as centers of circles that will overlap. Choose a radius (the distance between the point and pencil on the compass) that is larger than half the segment. Connect the points where the two circles intersect. This vertical line segment CD is perpendicular to the given segment AB and bisects the segment at point E.

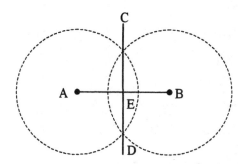

B. Bisect angle LMG.

Construct a circle using M as the center. The intersection points where the circle M and the angle meet are labeled points S and R. Construct two circles using points S and R as center points. Circles S and R overlap. Connect point M with the overlapping points, E and F, and get the bisector of angle LMG, which is segment EM.

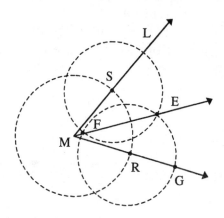

C. Duplicate angle QVJ.

Draw a ray with endpoint H. Construct a circle with V as the center such that circle V intersects the angle at points Y and K. Keep the same radius and draw a circle using the endpoint H of the ray as the center of the circle. Circle H intersects the ray at point Z. Adjust the radius of your compass to measure the distance between points Y and K. With this distance, place the point of your compass on Z and construct a circle. Circles H and Z intersect at point F. Construct the second ray with points H and F. Angle FHZ is an exact copy of angle QVJ.

D. Construct a triangle given three segments.

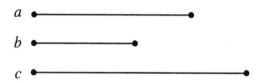

The plan is to duplicate one of the given segments. On the endpoints of this segment, construct a circle so that the radius of each circle is the length of one of the remaining segments. These two circles will overlap and the point of intersection will be the third vertex of the triangle. Draw a line segment. Mark a point on this segment as one of the endpoints of segment c. Find the distance of c using your compass. With the radius set, place the point of the compass on the point of the line segment and mark off the length of c with the pencil end. Use the left endpoint to construct a circle with a radius the length of segment a. Do the same with the right endpoint, but this circle has a radius the length of segment b. Use the intersection points of these two circles as the third vertex and construct the triangle.

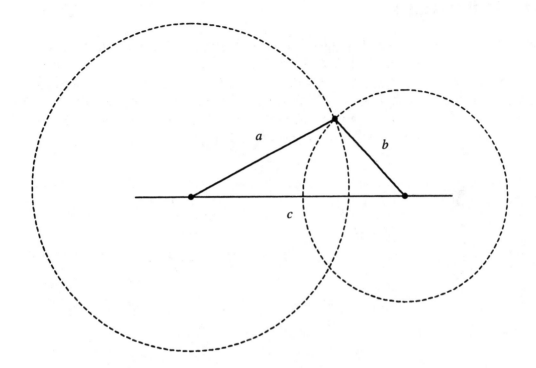

TEST YOUR SKILLS

1. Which line segments form angle *k*?

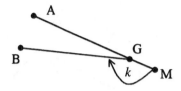

  A. $\overline{AM}$ and $\overline{BM}$
  B. $\overline{AG}$ and $\overline{BG}$
  C. $\overline{MG}$ and $\overline{BG}$
  D. $\overline{MG}$ and $\overline{AG}$

2. Given: $\overline{QT}$ = 10, $\overline{WT}$ = 18, $\overline{QW}$ = 12, and ∠Q is a right angle. Which of the following statements is true?

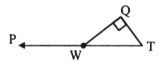

  A. Triangle WQT is an isosceles triangle.
  B. Angles QWT and QTW are supplementary.
  C. Angles PWT and QTW are supplementary.
  D. Angles QWT and QTW are complementary.

3. The lengths of three segments are 2, 7, and 9 centimeters. Using any segment as many times as you wish, answer the following questions.

   A. How many equilateral triangles can be constructed with these segments? Explain your answer.
   B. How many isosceles triangles can be constructed? Explain your answer.
   C. How many scalene triangles can be constructed? Explain your answer.

4. △ABC ~ △CDE and ∠ACD and ∠BCE are straight angles. Give two reasons why ∠a = ∠b.

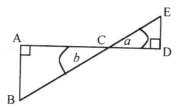

5. Construct a triangle using these two segments and this angle.

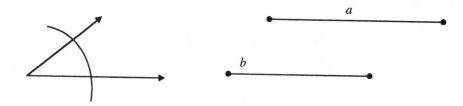

6. You are told that △ABC ~ △FED. Find the lengths of $\overline{DE}$ and $\overline{FD}$. Show your work.

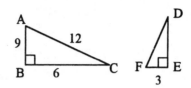

7. Choose the one correct statement and explain why it is true.

   A. All squares are rectangles.
   B. All trapezoids are parallelograms.
   C. All rectangles are squares.

8. Find the false statement and tell why it is false.

   A. All circles are similar to each other.
   B. If the sum of two angles is 90°, the angles are complementary.
   C. A chord is a diameter.
   D. All three angles in an equilateral triangle are congruent.

9. A square contains a circle. The radius of the circle is 2 feet. What is the relationship of the radius to the side of the square?

10. A pyramid has the top removed so the removed section is similar to the original pyramid. Then a vertical cut is made so that the plane of the cut and the base of the pyramid are perpendicular. What two-dimensional figure is on the vertical face?

TEST YOUR SKILLS SOLUTIONS

1. **C** is correct. Point G is shared by both segments that form ∠k. Another name for ∠k is ∠BGM.

2. **D** is correct. The sum of the interior angles of a triangle is 180°. Since ∠Q is 90°, the sum of ∠W and ∠T must be 90°. So the angles are complementary.

3. **A.** You can construct three triangles. One has sides of 2 cm, one has sides of 7 cm, and one has sides of 9 cm.

   B. There are six possible combinations of 2, 7, and 9, but only four triangles can be constructed. They can have sides of 2, 7, and 7 cm; 2, 9, and 9 cm; 7, 9, and 9 cm; and 9, 7, and 7 cm. You cannot construct triangles of 2, 2, and 7 or 2, 2, and 9 because the sum of any two sides of a triangle *must* be greater than the third side.

   C. You cannot construct a scalene triangle with these lengths. The sum of 2 and 7 is 9, which is the third side. The sum of any two sides of a triangle must be greater than the third side.

4. Reason #1: The measure of corresponding angles is equal in similar triangles.

   Reason #2: Angle *b* = ∠*a* because they are vertical angles. Vertical angles are always equal in measure.

5. Draw a line segment and duplicate segment *a*. Duplicate the angle so the vertex is one of the endpoints of segment *a*. Duplicate the length of segment *b* along the ray of the angle. Finish constructing the triangle by connecting the endpoints of segments *a* and *b*.

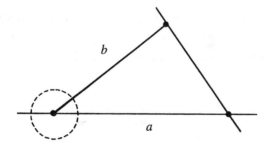

6. Rotate the triangles if you need to see the corresponding sides more clearly.

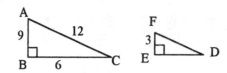

Pairs of corresponding sides in similar triangles have the same ratio.

$$\frac{AB}{FE} = \frac{BC}{ED} = \frac{AC}{FD}$$

To find $\overline{DE}$, set up a proportion.

$$\frac{AB}{FE} = \frac{BC}{ED}$$

Substitute the measure of the lengths for the sides.

$$\frac{9}{3} = \frac{6}{ED}$$

Cross multiply and get the length of $\overline{DE}$ as 2 units.

To find $\overline{FD}$, follow the same procedure using a proportion.

$$\frac{AB}{FE} = \frac{AC}{FD}$$

$$\frac{9}{3} = \frac{12}{FD}$$

Cross multiply and $\overline{FD}$ = 4 units.

7. The correct statement is **A**. A square has the same attributes of a rectangle, but it also has one more—all sides are congruent with each other. Statement **B** is false because trapezoids are not parallelograms; trapezoids have only one pair of parallel sides. Statement **C** is false because you can construct a rectangle that is not a square.

8. Statement **A** is true. Any two circles are similar because they have the same shape. Statement **B** is true by definition of complementary angles. Statement **D** is true for equilateral triangles. Statement **C** is false because not all chords are diameters. A diameter is a special kind of chord that passes through the center of the circle.

9. Remember that a radius is the distance between the center of a circle to *any* point on the circumference. Draw another radius from the center of the circle to a point on the circumference that touches the side of the square. A radius of 2 feet makes the diameter 4 feet. The diameter is the same length as the side of the square. So, the radius is half the length of the side of the square.

10. Picture a pyramid in your mind.

Then cut off the top of the pyramid.

Cut the figure in half vertically.

The figure on the vertical face is an isosceles trapezoid.

Chapter 6

Perimeter, Area, and Volume

PERIMETER

Think *distance around* when you see the word *perimeter*. When you are looking for the perimeter of a two-dimensional or plane figure, you are adding the lengths of the sides of the figure. Formulas help us do the arithmetic.

Here is the formula for the perimeter of a rectangle.

$$P_{rectangle} = 2(l + w) = 2l + 2w \quad (l = \text{length and } w = \text{width})$$

Example

A. Find the perimeter of a rectangle whose width is 3 cm and whose length is twice the width.

Draw a rectangle. Label the lengths: $w = 3$ and $l = 6$.

Now add the sides.

$$P = 2 (l + w)$$
$$P = 2 (3 + 6)$$
$$P = 2 (9)$$
$$P = 18 \text{ cm}$$

The formula for the perimeter of a square is similar to that of a rectangle. However, since the sides of the square have the same length, multiply the length of one side by 4. Multiplying by 4 gives you the same answer as adding the four numbers, but it is faster.

$$P_{\text{square}} = 4s \quad (s = \text{side})$$

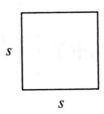

Example

B. Find the length of a side of a square whose perimeter is 64 mm.

Using the formula, substitute 64 for P.

$$P = 4s$$
$$64 = 4s$$
$$16 = s$$

The sides of the square are 16 mm each.
To find the perimeter of a triangle, add the lengths of the three sides together.

$$P_{\text{triangle}} = s_1 + s_2 + s_3$$

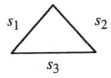

Example

C. An isosceles triangle has a base of 11 inches and a perimeter of 29 inches. What is the length of one of the congruent sides?

$P = s_1 + s_2 + s_3$ is the formula to use. Since two of the sides are the same length, we can rewrite the formula as $P = s_1 + 2s$ and then substitute the numbers.

$$P = s_1 + 2s$$
$$29 = 11 + 2s$$
$$18 = 2s$$
$$9 = s$$

The congruent sides measure 9 inches each.

Circles have different terminology. The perimeter of a circle is called its *circumference*. The formula used to calculate the circumference is

$$C = \pi d$$

Approximations for π are 3.14 or $\frac{22}{7}$. So the distance around a circle is a little more than three times its diameter.

Example

D. Find the circumference of a circle whose radius is 6 feet.

We know the diameter is twice the radius, or 12 feet. The answer can be represented in a few ways:

$$C = 12\pi \text{ feet when left in pi form}$$

$$C \approx 12(3.14) \text{ or } 37.68 \text{ feet or}$$

$$C \approx \frac{22}{7}(12) \text{ or } \frac{264}{7} \text{ or } 35\frac{5}{7} \text{ feet}$$

Notice that the symbol "≈" is used when approximating an answer and the symbol "=" is used when π is in the answer.

AREA

Think of *covering* a plane figure in small squares when finding the *area*. There are formulas to make working easier, but try to understand how the formulas evolved. Remember to label your answer in square units such as 14 in.^{2} or 14 square inches. You are working with two dimensions—length and width—so you need to use *square units* as part of your answer.

$$A_{\text{rectangle}} = lw$$

To find the area of a rectangle, simply multiply the length by the width.
In this case, $l = 9$ and $w = 4$. The area is 36 square units. It would take 36 squares to cover this rectangle.

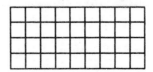

The formula for finding the area of a square is similar to the rectangle formula, except the length and width are the same measure.

$$A_{\text{square}} = s^2$$

This square has a side of 5. The area is 25 units^2.

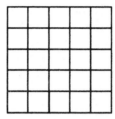

Here is the formula for finding the area of a triangle.

$$A_{\text{triangle}} = \frac{hb}{2} \quad \text{where } h = \text{height and } b = \text{base}$$

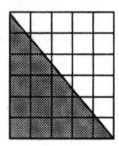

This triangle was formed by cutting a rectangle in half. The area of the rectangle is 30 units^2, so the area of the triangle is 15 units^2. The base of this triangle is 5 units, and the height is 6 units. *Any* triangle that has a base (or height) of 5 and height (or base) of 6 has an area of 15 units^2, no matter what the shape. All the following triangles have the same area.

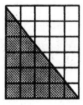 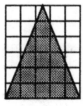 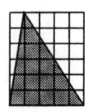 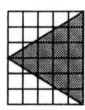

To find the area of a trapezoid, add the lengths of the parallel bases and multiply by the height. Then divide in half.

$$A_{\text{trapezoid}} = \frac{(b_1 + b_2)h}{2}$$

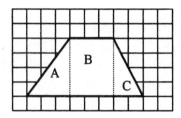

If you can't remember the formula, use your problem-solving techniques of dividing a big problem into smaller ones. Trapezoids can be divided into rectangles and triangles. Find the area of each smaller figure and then add up the numbers to get the area of the entire trapezoid.

Example

A. Find the area of this trapezoid. The scale is 1 unit = 5 mm.

Suppose that you forgot the formula for the area of a trapezoid but remember how to get the area of triangles and rectangles. Break the trapezoid into smaller regions and then find the areas of each. Add your three answers together to get the area of the entire trapezoid.

$$A_A = \frac{bh}{2} = \frac{(15)(20)}{2} = 150 \text{ mm}^2$$

$$A_B = lw = (15)(20) = 300 \text{ mm}^2$$

$$A_C = \frac{bh}{2} = \frac{(10)(20)}{2} = 100 \text{ mm}^2$$

$$A_{\text{trapezoid}} = 150 \text{ mm}^2 + 300 \text{ mm}^2 + 100 \text{ mm}^2 = 550 \text{ mm}^2$$

An important thing to remember is that the height (or altitude) of a figure is a perpendicular distance, *not* the length of a side. When you are measured for your height, you stand straight and tall—you don't lean!

Example

B. Find the area of trapezoid ABCD, where the height = 16 cm, the length of base AB = 10 cm, and the length of base CD = 40 cm.

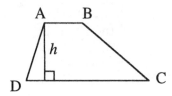

Using the formula, we substitute:

$$A_{\text{trapezoid}} = \frac{(b_1 + b_2)h}{2}$$

$$= \frac{(10 + 40)16}{2}$$

$$= 400 \text{ cm}^2$$

Here is the formula for finding the area of a circle.

$$A_{\text{circle}} = \pi r^2$$

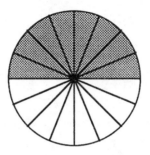

This circle is cut into 16 sections; half of them are shaded. If we cut this circle on the lines and arrange them into a rectangle we get this figure.

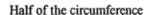

Half of the circumference

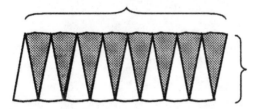

The radius

The smaller the pie-shaped pieces, the closer the figure gets to looking like a rectangle. The length of this rectangle is half the length of the circumference and the height is equal to the radius of the circle. We know that the circumference is equal to πd, and we represent the radius as r. The area of this rectangle is

$$\frac{\pi d}{2} \times r \;\; = \;\; \frac{2\pi r}{2} \times r = \frac{\cancel{2}\pi r}{\cancel{2}} \times r$$

$$= \pi r \times r = \pi r^2$$

So, the formula for the area of a circle is $A_{\text{circle}} = \pi r^2$.

Examples

C. Find the area of a circle whose radius is 4 cm. Keep your answer in pi form.

Using the formula, substitute 4 for r.

$$A = \pi r^2$$

$$A = \pi(4^2)$$

$$A = 16\pi \text{ sq cm} = 16\pi \text{ cm}^2$$

Remember to label your answer with *square units* because this is *area*. Also notice that if your answer contains π, then the number precedes the symbol. In this case, 16 is written before π.

D. Find the area of a circle with a diameter of 20 inches.

If the diameter is 20 inches, then the radius is 10.

$$A = \pi r^2$$

$$A \approx (3.14)(10^2)$$

$$A \approx (3.14)(100)$$

$$A \approx 314 \text{ in.}^2$$

E. You have a rectangular piece of wood that measures 24 inches by 15 inches. To the nearest inch, find the area of the largest circle you can get from this piece of wood.

The largest diameter will be 15 inches. So find the area of a circle with a 15-inch diameter.

$$A = \pi r^2$$

$$A \approx (3.14)(7.5^2)$$

$$A \approx (3.14)(56.25)$$

$$A \approx 176.63 \text{ in.}^2$$

$$A \approx 177 \text{ in.}^2$$

SURFACE AREA

You may have to find the surface area of a three-dimensional object. Don't get surface area confused with volume. You are still looking for the amount of material to cover an object. The idea behind finding the surface area is to find the area of each face of a three-dimensional object and then to add the areas together for your final answer.

The surface area of a sphere has a special formula:

$$SA = 4\pi r^2$$

Examples

A. How much contact paper will you need to cover a box in the diagram?

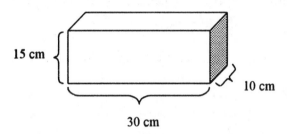

You will be covering six rectangles, two each of three different dimensions. Find the area of the three different rectangles and then double the area.

1. The top and bottom rectangles are 10 cm by 30 cm.
2. The front and back rectangles are 15 cm by 30 cm.
3. The side rectangles are 10 cm by 15 cm.

$$A_{top} = 300 \text{ cm}^2$$

$$A_{front} = 450 \text{ cm}^2$$

$$A_{side} = 150 \text{ cm}^2$$

The sum is 900 cm^2 for three of the rectangles or 1800 cm^2 for the entire box.

B. The Kindergarten teacher asked you to paint a box of wooden blocks to give them a new look. The box contains 10 blocks of three different sizes, or 30 blocks. The teacher wants you to paint the small blocks red, the medium blocks yellow, and the large blocks blue. The blocks are all cubes: the small ones have an edge of 3 inches, the medium of 6 inches, and the large of 10 inches. A pint of paint will cover 5 square feet. She gave you 2 pints of each color. Will you have enough paint?

The area of one face of the small block is 9 in.^{2}. There are six faces. So the surface area of one small block is 54 in.^{2}. Ten of them need to be painted. You will need to cover 540 square inches in red paint. One can of paint will cover 5 square feet. There are 144 square inches in 1 square foot. In 5 square feet, there are 720 square inches. Yes, there is enough red paint for the small blocks.

The area of one face of the medium block is 36 square inches or 216 square inches for the entire block. You have to cover 2160 square inches of the medium blocks in yellow. You know that one can will cover 720 square inches, so two cans can cover 1440 square inches. You need to cover 720 square inches more (2160 − 1440 = 720), so you need to buy a 1-pint can of yellow paint.

The surface area of one large block is 600 square inches. To cover ten blocks you need enough paint to cover 6000 square inches. If 1 pint covers 720 in.^{2}, divide 6000 by 720 and see that you need about 8.33 pints or 9 pints to paint all the blocks. You need to get 7 pints of blue paint. (Remember that a gallon contains 8 pints. This may be another option.)

C. Find the approximate surface area of this can to the nearest square inch where the circumference of the base is about 6.25 inches and the height is 5.5 inches.

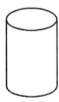

The surface area will be the sum of the two bases (circles) and the cylinder (rectangle).

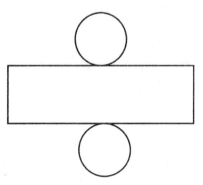

If you cut the can so that it lies flat, you get this object.

$$C = \pi d$$

$$6.25 = \pi d$$

$$\frac{6.25}{3.14} = d$$

$$d \approx 1.99$$

$$r \approx 1$$

$$A_{\text{circle}} = \pi r^2$$

$$A_{\text{circle}} = \pi(1^2)$$

$$A_{\text{circle}} \approx 3.14$$

Double this area and get 6.28 in.^{2} for the area of both bases. The length of the rectangle is the circumference of the circle.

$$A_{\text{rectangle}} = lw$$

$$A_{\text{rectangle}} = (6.25)(5.5)$$

$$A_{\text{rectangle}} = 34.375 \text{ in.}^2$$

$$SA = 34.375 + 6.28$$

$$SA = 40.655 \text{ or } 41 \text{ in.}^2$$

VOLUME

Think of *filling* a container with cubes when finding the *volume*. Find the area of the base of the object and then multiply by the height to get the volume. The following formulas make our work easier.

$$V_{\text{any rectangular solid}} = lwh \quad (\text{length} \times \text{width} \times \text{height})$$

$$V_{\text{cube}} = e^3 \quad (e \text{ represents the length of an edge. All three dimensions are the same in a cube, which is a special instance of a rectangular solid.})$$

$$V_{\text{cylinder}} = \pi r^2 h$$

$$V_{\text{sphere}} = \frac{4}{3}\pi r^3$$

$$V_{\text{cone}} = \frac{\pi r^2 h}{3}$$

TEST YOUR SKILLS

1. Find the area of a circle whose diameter is 2.5 cm. Use 3.14 as an approximation for pi. Round your answer to the nearest tenth. Show your work.

2. Find the length of the side of the square whose perimeter and area have the same number. Explain your answer.

3. Find the area of this trapezoid in square millimeters.

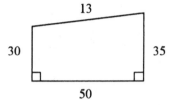

4. Find the surface area of a cube that has an edge of 8 inches. Show your work.

5. A square has an area of 256 square meters. To the nearest whole number, what would the radius of a circle be so that the circle would have approximately the same area?

6. Find the ratio of the area of the smaller square to the area of the larger square in the diagram.

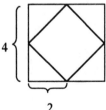

7. The stage floor in the school auditorium needs two coats of paint. The semicircular section of the stage is centered on the rectangular section. How much paint is needed if 1 gallon covers 80 square feet in one coat?

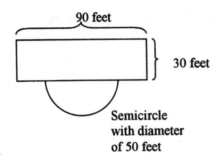

90 feet

30 feet

Semicircle
with diameter
of 50 feet

8. A triangle has an area of 6 square feet. What could the height and length be in this triangle? List as many integer values as you can.

9. The volume of a certain cube is close in measure to the volume of a sphere with a radius of 6 cm. Find the edge of this cube to the nearest tenth.

10. Your neighbor asked you to help her enclose a garden in her yard. The length of the rectangular garden will be twice the width. A roll of fencing is 100 feet long. What are the whole-number dimensions of the largest garden that can be enclosed?

TEST YOUR SKILLS SOLUTIONS

1. Divide the diameter in half to get a radius of 1.25 cm. Use the formula.

$$A = \pi r^2$$

$$A \approx (3.14)(1.25^2)$$

$$A \approx (3.14)(1.5625)$$

$$A \approx 4.90625$$

$$A \approx 4.9 \text{ cm}^2$$

2. If the perimeter is equal to the area, then the formulas are equal.

Write the equation.

$$4s = s^2$$

Write the factors separately.

$$4 \times s = s \times s$$

Divide both sides by s.

$$4 = s$$

Thus, the side of the square is 4 units. So, the perimeter is 16 units and the area is 16 units^2.

3. Use the formula. The bases are parallel in a trapezoid. The parallel sides are 30 and 35. Be careful in identifying the parts.

$$A = \frac{(b_1 + b_2)h}{2}$$

$$A = \frac{(30 + 35)(50)}{2}$$

$$A = \frac{(65)(50)}{2}$$

$$A = 1625 \text{ mm}^2$$

4. Each of the six faces of a cube is identical. Find the area of one face and multiply by 6 to get the surface area. The area of one face is 64 in.^{2} ($A = s^2$).

$$64 \times 6 = 384$$

The surface area of the cube is 384 in.^{2}

5. The problem says a circle and square have approximately the same area. Find the radius of the circle by substituting what you know into the formula.

$$A = \pi r^2$$

$$256 \approx (3.14)(r^2)$$

$$\frac{256}{3.14} \approx \frac{(3.14)r^2}{3.14}$$

$$81.53 \approx r^2$$

$$\sqrt{81.53} \approx \sqrt{r^2}$$

$$9 \approx r$$

The radius of the circle is approximately 9 square meters. If you substitute 9 for the radius in the area formula, you see that $(9^2)(3.14)$ is approximately 254.3, which is very close to 256.

6. There is more than one way to solve a problem. To find the area of the smaller square, cut the square into 4 triangles, then rearrange them.

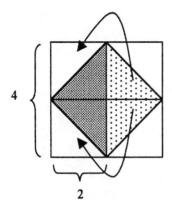 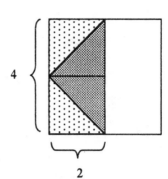

The four triangles form a rectangle that has an area of 8 units^2. The area of the larger square is 16 units^2. The ratio of smaller to larger is 1 to 2.

7. Take the area of the semicircle and the rectangle and add them together.

$$A = \frac{\pi r^2}{2} \qquad\qquad A = lw$$

$$A \approx \frac{(3.14)(25^2)}{2} \qquad A = (90)(30)$$

$$A \approx \frac{1963}{2} \qquad\qquad A = 2700 \text{ ft}^2$$

$$A \approx 982 \text{ ft}^2$$

Together, the area of the stage is 982 + 2700 or approximately 3682 ft^2. If 1 gallon covers 80 ft^2 in one coat, it will cover 40 ft^2 in two coats. Divide 3682 by 40 and get 92.05, which is the number of gallons needed. Round up your answer to buy 93 gallons of paint.

8. The formula for the area of a triangle is

$$A = \frac{bh}{2}$$

Substitute 6 for A

$$6 = \frac{bh}{2}$$

Multiply both sides by 2

$$(2)6 = \frac{bh}{2}(2)$$

$$12 = bh$$

List the pairs of factors that result in 12: (1, 12), (2, 6), and (3, 4). Any of these combinations of base and height will give you a triangle whose area is 6 ft^2.

9. First find the volume of the sphere where $r = 6$ cm.

$$V = \frac{4\pi r^3}{3}$$

$$V \approx \frac{4(3.14)(6^3)}{3}$$

$$V \approx \frac{4(3.14)(216)}{3}$$

$$V \approx \frac{2712.96}{3}$$

$$V \approx 904.32 \text{ cm}^3$$

This is also the approximate volume of the cube. Substitute this value into the formula.

$$V = e^3$$

$$904.32 \approx e^3$$

$$\sqrt[3]{904.32} \approx \sqrt[3]{e^3}$$

$$9.6703 \approx e$$

To the nearest tenth, the length of the edge of the cube is 9.7 cm.

10. $P = 2(l + w)$ is the formula for the perimeter of a rectangle. The problem states that the length is twice the width. Substitute $2w$ for l in the formula. That way there is only one variable to work with. The perimeter is about 100 feet.

$P = 2(l + w)$	Use the perimeter formula.
$100 = 2(2w + w)$	Substitute $2w$ for l.
$100 = 2(3w)$	Combine like terms.
$100 = 6w$	Divide both sides by 6.
$16.7 \approx w$	

You cannot go over 100 feet. If we round up 16.7 to 17, the perimeter of the garden is 2(17 + 34) or 2(51), which will put us over the 100-feet maximum. If we try truncating our answer the perimeter will be 2(16 + 32) or 2(48) or 96 feet, which is close to, yet under, 100 feet. Thus, the dimensions of the garden are 16 feet by 32 feet.

Chapter 7

Trigonometry

The study of the relationship between the angles and the sides of a right triangle is called trigonometry. This word has Greek origins that mean triangle measurement.

The variable x is used in algebra to represent an unknown quantity. When representing an unknown measure of an angle, some people use the Greek letter θ, or *theta* (pronounced THAY-tuh). Only right triangles are studied in trigonometry.

PYTHAGOREAN THEOREM

Pythagoras was a Greek philosopher (c. 560–c. 480 B.C.E.). He headed a secret society, which had many followers that valued knowledge highly. Some say that Pythagoras did not discover the theorem that was credited to him but rather that one of his followers discovered it. Whatever the case, the Pythagorean Theorem is still used today in our world.

If you square the two legs of a right triangle, the square of the hypotenuse is equal to their sum. We know this as $a^2 + b^2 = c^2$, or the Pythagorean Theorem.

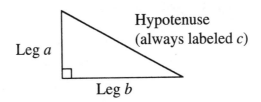

Leg a

Hypotenuse
(always labeled c)

Leg b

Example

A. Find the length of the hypotenuse of a triangle whose legs are 3 and 4.

$$a^2 + b^2 = c^2$$

$$3^2 + 4^2 = c^2$$

$$9 + 16 = c^2$$

$$25 = c^2$$

$$c = +5 \text{ or } -5$$

Since lengths are not negative, we disregard −5. Consider only the positive value of the square root for the rest of this chapter. The length of the hypotenuse is 5.

This grouping (3-4-5) is called a *Pythagorean triple*. Multiples of 3-4-5 are also triples, such as 9-12-15 and 15-20-25.

Example

B. The hypotenuse of a triangle is 10, and one of its legs is 6. What is the length of the other leg?

Substitute the values into the formula.

$$a^2 + b^2 = c^2$$

$$a^2 + 6^2 = 10^2$$

$$a^2 + 36 = 100$$

$$a^2 = 64$$

$$a = 8$$

The length of the other leg is 8.

You may have noticed that 6-8-10 is a triple. It is a multiple of 3-4-5. Two other triples that are good to know are 5-12-13 and 7-24-25.

Examples

C. Find the length of the hypotenuse when the legs have lengths of 7 and 10.

$$a^2 + b^2 = c^2$$

$$7^2 + 10^2 = c^2$$

$$49 + 100 = c^2$$

$$149 = c^2$$

$$c = \sqrt{149}$$

$\sqrt{149}$ is not a perfect square. You can leave your answer in this form or approximate it.

$$\sqrt{144} < \sqrt{149} < \sqrt{169}$$

$$12 < \sqrt{149} < 13$$

The $\sqrt{149}$ is between 12 and 13. You can approximate your answer to 12.2 with your calculator.

D. The hypotenuse of a triangle is 35. One leg has a length of 20. Find the length of the other leg to the nearest tenth.

$$a^2 + b^2 = c^2$$

$$a^2 + 20^2 = 35^2$$

$$a^2 + 400 = 1225$$

$$a^2 = 825$$

$$a = \sqrt{825}$$

$$a \approx 28.7$$

TRIGONOMETRIC FUNCTIONS

Sine, cosine, and tangent are three of the trig functions that will be discussed here. In high school you will learn three more.

Each of these functions is a ratio between the sides and the hypotenuse.

$$\sin \theta = \frac{\text{Opposite}}{\text{Hypotenuse}} \qquad \left(\frac{\text{Oscar}}{\text{had}} \right)$$

$$\cos \theta = \frac{\text{Adjacent}}{\text{Hypotenuse}} \qquad \left(\frac{\text{a}}{\text{heap}} \right)$$

$$\tan \theta = \frac{\text{Opposite}}{\text{Adjacent}} \qquad \left(\frac{\text{of}}{\text{apples}} \right)$$

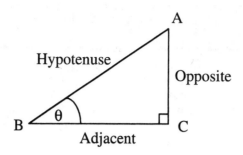

Side BC is adjacent to ∠θ because BC is one of the segments that form the angle. Side AC is opposite because it does not help form the angle. The hypotenuse is *always* the longest side.

Sin is the abbreviation for sine just as *cos* is the abbreviation for cosine and *tan* stands for tangent. There are mnemonics to help remember the ratios. "Oscar had—a heap—of apples" is an old one.

Angle BAC or angle A will be referred to as θ. Now we will find the measure of θ using our ratios and the trigonometric function table. This table will be provided in the test booklet.

Examples

A. Find the measure of ∠θ to the nearest degree.

We know two measurements—the opposite and the hypotenuse. To find θ, we will use sine.

$$\sin \theta = \frac{o}{h} = \frac{12}{20} = \frac{3}{5} = 0.6$$

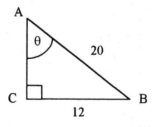

Look in the table under *Sin* and find 0.6. This is the closest to 0.6018. Angle θ is 37° to the nearest degree.

Trigonometric Table

Angle	Sine	Cosine	Tangent	Cotangent	
0°	.0000	1.000	.0000	—	90°
1°	.0175	.9998	.0175	57.290	89°
2°	.0349	.9994	.0349	28.636	88°
3°	.0523	.9986	.0524	19.081	87°
4°	.0698	.9976	.0699	14.301	86°
5°	.0872	.9962	.0875	11.430	85°
6°	.1045	.9945	.1051	9.5144	84°
7°	.1219	.9925	.1228	8.1443	83°
8°	.1392	.9903	.1405	7.1154	82°
9°	.1564	.9877	.1584	6.3138	81°
10°	.1736	.9848	.1763	5.6713	80°
11°	.1908	.9816	.1944	5.1446	79°
12°	.2079	.9781	.2126	4.7046	78°
13°	.2250	.9744	.2309	4.3315	77°
14°	.2419	.9703	.2493	4.0108	76°
15°	.2588	.9659	.2679	3.7321	75°
16°	.2756	.9613	.2867	3.4874	74°
17°	.2924	.9563	.3057	3.2709	73°
18°	.3090	.9511	.3249	3.0777	72°
19°	.3256	.9455	.3443	2.9042	71°
20°	.3420	.9397	.3640	2.7475	70°
21°	.3584	.9336	.3839	2.6051	69°
22°	.3746	.9272	.4040	2.4751	68°
23°	.3907	.9205	.4245	2.3559	67°
24°	.4067	.9135	.4452	2.2460	66°
25°	.4226	.9063	.4663	2.1445	65°
26°	.4384	.8988	.4877	2.0503	64°
27°	.4540	.8910	.5095	1.9626	63°
28°	.4695	.8829	.5317	1.8807	62°
29°	.4848	.8746	.5543	1.8040	61°
30°	.5000	.8660	.5774	1.7321	60°
31°	.5150	.8572	.6009	1.6643	59°
32°	.5229	.8480	.6249	1.6003	58°
33°	.5446	.8387	.6494	1.5399	57°
34°	.5592	.8290	.6745	1.4826	56°
35°	.5736	.8192	.7002	1.4281	55°
36°	.5878	.8090	.7002	1.3764	54°
37°	.6018	.7986	.7536	1.3270	53°
38°	.6157	.7880	.7813	1.2799	52°
39°	.6293	.7771	.8098	1.2349	51°
40°	.6428	.7660	.8391	1.1918	50°
41°	.6561	.7547	.8693	1.1504	49°
42°	.6691	.7431	.9004	1.1106	48°
43°	.6820	.7314	.9325	1.0724	47°
44°	.6947	.7193	.9657	1.0355	46°
45°	.7071	.7071	1.0000	1.0000	45°
	Cosine	Sine	Cotangent	Tangent	Angle

B. Find the measure of ∠CBA.

You know the lengths of all the sides. You have a choice in which function to use. Use tangent.

$$\tan \theta = \frac{o}{a} = \frac{20}{48} = \frac{5}{12}$$

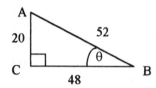

Use your scientific calculator to find the measure of ∠θ in degrees. Change the fraction into a decimal by dividing.

$$5 \div 12 \approx 0.4166667$$

With your answer in the calculator, press the "2nd" button and then press the "TAN" button. Your answer rounded to the nearest tenth of a degree is 22.6°.

C. Using the diagram in the preceding example, find the measure of ∠θ using the cosine function.

$$\cos \theta = \frac{a}{b} = \frac{48}{52} = \frac{12}{13} \approx 0.9231$$

Using your calculator, find the degrees to the nearest tenth. Press the "2nd" button and then press the "COS" button. You should get 22.6°. As you can see, the answer is the same as in the previous example.

D. Using the same diagram again, find the measure of ∠θ using the sine function.

$$\sin \theta = \frac{a}{b} = \frac{20}{52} = \frac{5}{13} \approx 0.3846 \approx 67.4°$$

This answer should make sense to you because the acute angles of a right triangle are complementary. The sum of 22.6° and 67.4° is 90°.

TEST YOUR SKILLS

1. The owner of an apartment building needs to repair the roof. The building is 40 feet tall, and he has the bottom of the ladder 20 feet away from the building. What is the measure of the angle between the base of the ladder and the ground? Show your work.

2. Tim is in his backyard at Y and his house is at X. How much shorter is his path along the diagonal compared to the walk along the sides of the rectangle?

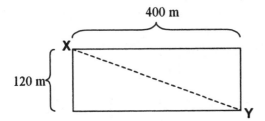

3. A right isosceles triangle has sides of length 5 meters each. Find the length of the base to the nearest tenth. Show your work.

4. A 15-foot ladder is 6 feet away from the wall. How far up does the ladder touch the wall? Show your work.

5. Write the ratios for sine, cosine, and tangent used to find the measure of angle A.

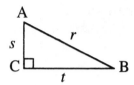

6. Find the length of the diagonal of an 8½ by 11-inch sheet of paper using the Pythagorean Theorem.

7. The bottom of a rectangular pencil box measures 20 mm by 10 mm. To the nearest whole millimeter, what is the length of the longest pencil that will fit in this box? Show your work.

8. Find the perimeter of this triangle to the nearest inch given that $\overline{QS}$ = 14, $\overline{TS}$ = 10, and $\overline{RT}$ = 8. Show your work.

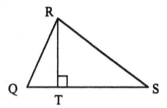

9. On a graph, plot point A at (2, 3) and point B at (5, 1). Find the distance between these points using what you know about right triangles. Explain your answer.

10. An equilateral triangle has a perimeter of 78 cm. What is the area of this triangle to the nearest centimeter? Show your work.

TEST YOUR SKILLS SOLUTIONS

1. The information you are given are the adjacent and opposite lengths. This means that you use the tangent to get the angle measure.

$$\tan\theta = \frac{o}{a} = \frac{40}{20} = 2$$

By using the "2nd" and "TAN" buttons on your calculator, you should find that the angle is 63.4°.

2.

$$a^2 + b^2 = c^2$$

$$120^2 + 400^2 = c^2$$

$$14{,}400 + 160{,}000 = c^2$$

$$174{,}400 = c^2$$

$$c \approx 418$$

The length of the shorter path is 418 meters. He would normally have to walk 400 + 120 or 520 meters. The diagonal path is about 102 meters shorter.

3. The base is the hypotenuse in this triangle.

$$a^2 + b^2 = c^2$$

$$5^2 + 5^2 = c^2$$

$$25 + 25 = c^2$$

$$50 = c^2$$

$$c \approx 7.1 \text{ m}$$

The base is about 7.1 meters.

4. In this problem, we know the hypotenuse and one of the legs but must find the other leg.

$$a^2 + b^2 = c^2$$

$$a^2 + 6^2 = 15^2$$

$$a^2 + 36 = 225$$

$$a^2 = 189$$

$$a \approx 13.7$$

So the ladder touches the wall about 14 feet from the ground.

5. Write the ratios with the variables r, s, and t.

$$\sin A = \frac{o}{h} = \frac{t}{r}$$

$$\cos A = \frac{a}{h} = \frac{s}{r}$$

$$\tan A = \frac{o}{a} = \frac{t}{s}$$

6.

$$a^2 + b^2 = c^2$$

$$8.5^2 + 11^2 = c^2$$

$$72.25 + 121 = c^2$$

$$c^2 = 193.25$$

$$c \approx 13.9 \text{ inches}$$

7.

$$a^2 + b^2 = c^2$$

$$10^2 + 20^2 = c^2$$

$$100 + 400 = c^2$$

$$c^2 = 500$$

$$c \approx 22.36$$

To the nearest millimeter, a 22-mm pencil will fit in the box.

8. Write in the lengths that were given in the problem. Then use the Pythagorean Theorem to find the lengths of the sides QR and RS.

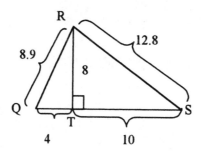

$a^2 + b^2 = c^2$	$a^2 + b^2 = c^2$
$8^2 + 4^2 = QR^2$	$8^2 + 10^2 = RS^2$
$64 + 16 = QR^2$	$64 + 100 = RS^2$
$QR^2 = 80$	$RS^2 = 164$
$QR \approx 8.9$	$RS \approx 12.8$

To find the perimeter, add the sides together (8.9 + 12.8 + 14) and get 35.7 inches. To the nearest inch, the perimeter of this triangle is 36 inches.

9. Plot the points. If you draw in the triangle, the segment with these two points as endpoints is the hypotenuse of a right triangle where the legs have lengths of 2 and 3. Use the Pythagorean Theorem.

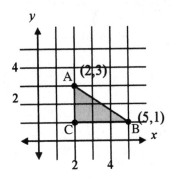

$$a^2 + b^2 = c^2$$

$$2^2 + 3^2 = c^2$$

$$4 + 9 = c^2$$

$$c^2 = 13$$

$$c \approx 3.6$$

10. All three sides of an equilateral triangle are the same. You need to find the height to get the area. Dropping an altitude bisects the base. Use the Pythagorean Theorem to get the height; then use the formula for the area of a triangle to get your final answer.

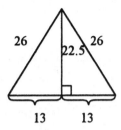

$$a^2 + b^2 = c^2 \qquad\qquad A = \frac{bh}{2}$$

$$13^2 + b^2 = 26^2 \qquad\qquad A \approx \frac{(26)(22.5)}{2}$$

$$169 + b^2 = 676$$

$$b^2 = 507 \qquad\qquad A \approx \frac{585}{2}$$

$$b \approx 22.5 \qquad\qquad A \approx 292.5 \text{ cm}^2$$

The area of the triangle is approximately 292.5 cm².

Chapter 8

Transformations

A transformation takes the points of a plane figure and moves them to another location on the plane according to a particular rule. This new figure is usually congruent (an exact copy) to the original. The exception is an image of a figure under a dilation. *Translations, reflections, rotations,* and *dilations* are types of transformations. Transformations are used in computer games.

TRANSLATION

Triangle ABC was translated to another location on the plane. Each point in the triangle has been shifted 7 units to the left. A translation is the same as sliding the figure to a new location on the plane.

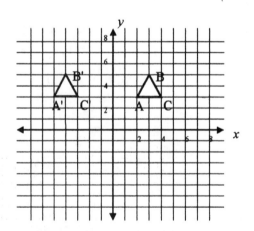

Example

A. The parallelogram on page 95 has the following coordinates: A = (6, 6), B = (7, 8), C = (8, 6), and D = (7, 4). After a translation, the coordinates are (1, 0) for A′ and (2, −2) for D′. What are the coordinates for points B′ and C′?

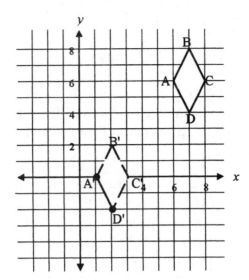

Plot points A′ and D′. Compare the location of segments AD and A′D′. The only possible location of B′ is (2, 2) and C′ is (3, 0) because the shape is not distorted in a translation.

Triangle KLM was translated vertically. Compare the coordinates of the original triangle and its image.

△KLM	△K′L′M′
K (2, –4)	K′ (2, 2)
L (2, –2)	L′ (2, 4)
M (6, –4)	M′ (6, 2)

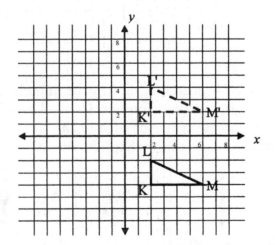

Notice that there is no change in the x-values of the points. The difference between the y-values of the original point at its image is 6 for all three points. Translating a figure horizontally will show that there is no change in the y-axis coordinates.

Example

B. Translate △K′L′M′ horizontally 7 units to the left. List the coordinates of the vertices.

In the diagram, there was no change in *x*-values for the vertical translation. There will be no change in the *y*-values for a horizontal translation.

△K′L′M′	△K″L″M″
K′ (2, 2)	K″ (−5, 2)
L′ (2, 4)	L″ (−5, 4)
M′ (6, 2)	M″ (−1, 2)

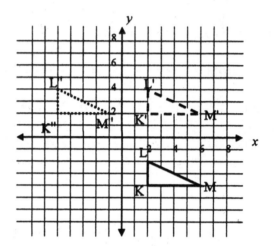

REFLECTION

Before going into the topic of reflection, recall your knowledge of symmetry. What do the letters H, T, M, Y, and A have in common? They all have a vertical line of symmetry. Draw a vertical line through the center of each of the letters. The shape on one side of the vertical line is a mirror image of the shape on the other side.

Example

A. How many letters in our alphabet can you find that have only a horizontal line symmetry?

B, C, D, E, and K.

Example

B. Do any letters have two lines of symmetry? Do any have more than two lines of symmetry?

H, I, and X have two while the letter O has an infinite number of lines of symmetry.

Example

C. How many lines of symmetry does a square have?

Four lines—vertical, horizontal, and along the two diagonals.

In a reflection, the figure is reflected in a line on the plane. If you traced the figure on paper and cut it out, the new location of the figure would be "flipped" over the line of reflection. Triangle ABC, when reflected over the *x*-axis results in triangle A′B′C′. The corresponding vertices are the same distance from the *x*-axis.

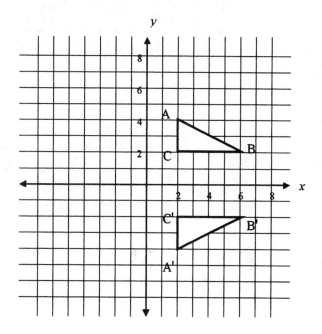

Examples

D. Reflect trapezoid WXYZ over the line $y = 7$.

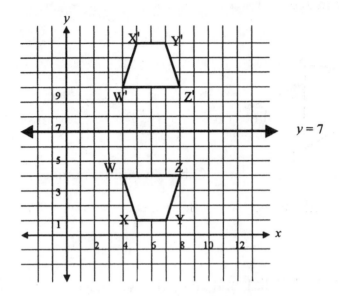

Identify the line of reflection. Count the distance point W is from the line $y = 7$. Plot the point W'. Do the same for the other three points. Draw in the line segments that form the new trapezoid W'X'Y'Z'.

E. Draw the image of triangle KPG under a reflection in line m.

Drop perpendicular lines from each vertex to line m. The location of each image point is determined by the distance the original point is from m. After all three new points are located, construct the triangle.

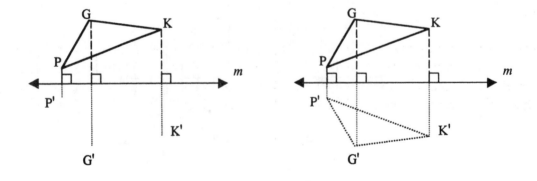

F. Draw the image of quadrilateral WXYZ under a reflection in line *l*.

Follow the same procedure as in Example E. When you have the location of the image points, construct the image.

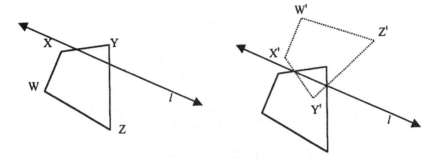

G. Which of the words below have reflective symmetry?

TAKE   BOOK   HI   SOON   CHICK

Check for horizontal line symmetry.

If the portion of the letter below the line is the mirror image of the portion above, then it is symmetrical. All the words except for TAKE and SOON have horizontal line symmetry.

H. Construct the line of reflection for these triangles.

Construct line segments connecting the corresponding vertices. Find the midpoints of each of the line segments because both triangles must be the same distance away from the line of reflection. The line you get by connecting the midpoints will be the line of reflection. This line is perpendicular to the line segments.

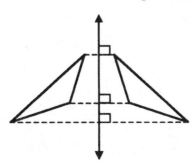

ROTATION

A rotation is accomplished when a figure has been turned about a fixed point on the plane. The letters N, S, and Z have rotational symmetry when the fixed point is in the center of the letter. The images are also congruent to the original figure after a turn.

N   S   Z

Another example of rotation symmetry uses the picture cards in a standard deck. Place your finger in the center of the Queen of Diamonds, for example. Turn the card 180°. The image you see now is the same as it was before you turned the card upside down.

Examples

A. Draw the flag of the image in the proper position after a rotation.

The pointed tip of the flag is to the right. The flag of the image will be pointed to the right also.

B. The hands of a clock now show 3:20. What time was it before the minute hand rotated 72°?

There are 60 minutes in one hour and 360° in a circle. So, each minute on a clock covers 6°. Divide 72° by 6° and get 12, which is the number of minutes the minute hand traveled over 72°. So, 12 minutes before 3:20 was 3:08.

DILATION

Just as the pupil of your eye will shrink or enlarge to adjust to the light, a geometric figure is stretched or shrunk under a dilation. In a dilation, the measure of each angle is preserved.

Examples

   A. Enlarge this right triangle so that its sides are doubled in length.

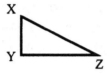

   Extend $\overline{YX}$ so that its length is doubled. The image of X is X'. Do the same for $\overline{YZ}$. Connect X' to Z'. Triangle X'Y'Z' is the image of the original triangle XYZ.

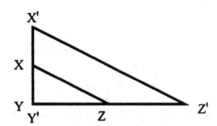

   B. Reduce the rectangle so its sides are one fifth of the original.

   Divide the lengths of the sides by five. The image of the original rectangle has lengths of 3 cm and 2 cm.

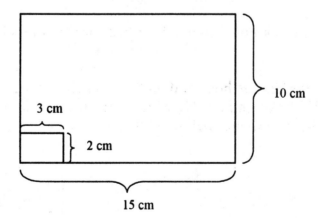

TEST YOUR SKILLS

1. Which of these figures has rotation symmetry? Explain your answer.

2. Explain how each of the figures B, C, and D are related to figure A.

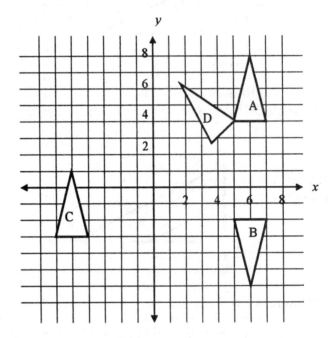

3. Which lines are lines of reflection? Explain your answer.

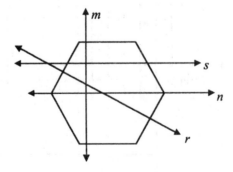

4. Reflect the quadrilateral over the *y*-axis and then translate this image down six units. List the coordinates of the new vertices.

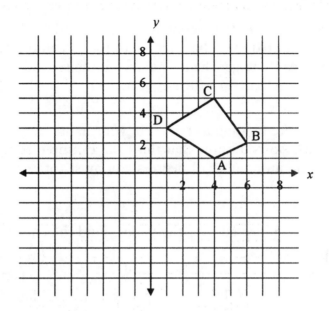

5. Explain how figure A was transformed to become figure B using translations, rotations, or reflections.

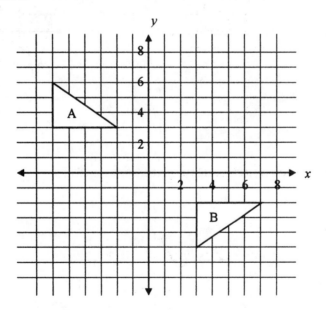

6. Explain why a rectangle has only two lines of symmetry and not four as does a square.

7. A right scalene triangle in the second quadrant is reflected over the y-axis. This new triangle is then reflected over the *x*-axis into the fourth quadrant. Instead of two transformations, can you get the triangle from the second to the fourth quadrant in one transformation? Explain your answer.

8. Suppose that you are holding a starfish in your hand. You notice that all five of its arms are identical, so you know that this is an example of rotational symmetry. What is the minimum number of degrees you must rotate the starfish to show this rotational symmetry?

9. The point (–4, 3) is reflected over the y-axis. Find the coordinates of the image point without using a graph.

10. Part of a figure is given in the following diagrams. The lines are lines of reflection. Complete each figure.

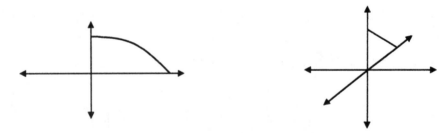

TEST YOUR SKILLS SOLUTIONS

1. This figure has horizontal- and vertical-line reflection symmetry.

This right scalene triangle has no symmetry.

This star has rotational symmetry. It can be rotated every 72°.

This figure has horizontal-line reflection symmetry.

2. Figure B is the image of A reflected over the line $y = 1$. Figure C is the image of A after a translation. Figure D is the rotated image of A.

3. Examine each line separately. Lines n and r cut the figure in half so that the shape on one side of the line is a mirror image of that on the other side.

4. Quadrilateral ABCD is reflected over the y-axis and is shown by a dotted line figure A′B′C′D′. The image translated down six units has coordinates of A″ (–4, –5), B″ (–6, –4), C″ (–4, –1), and D″ (–1, –3).

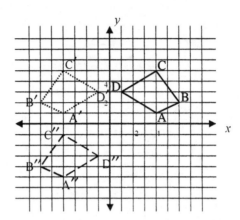

5. This can be accomplished in many ways. Figure A could be reflected over the line y = 3 to become figure A-1, then translated down 5 units to A-2, and finally translated 9 units to the right to B. Another way you can do this is to rotate A 180° to figure B-1, translate B-1 five units down to B-2, then reflect B-2 over the line y = 2.5 and get B.

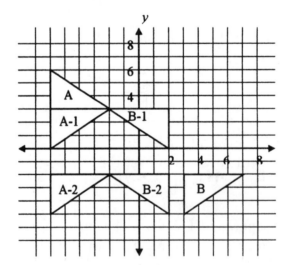

6. A rectangle has horizontal- and vertical-line reflection symmetry. It does not have symmetry along its diagonals. If you reflected the lower region of the rectangle over its diagonal, you get the following *kite*-shaped object.

7. A rotation of 180° will transform the triangle in one step. The point of rotation is the origin.

8. You would have to turn the starfish one-fifth of a circle. There are 360° in a circle and one-fifth of 360 is 72. The minimum number of degrees is 72.

9. The given point (–4, 3) has its image at (4, 3). A reflection over the y-axis does not change the value of the y-coordinate. The x-coordinate, –4, is 4 units to the left of the y-axis. The image of –4 is 4 units to the right of the y-axis at +4.

10. Here are the figures.

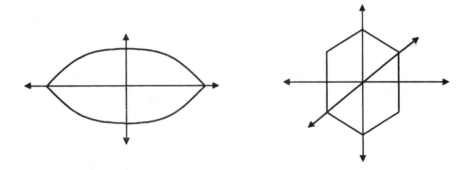

Chapter 9

Equations and Problem Solving

An *equation* is a mathematical sentence containing an equal sign and is constructed with terms. A *term* can be a variable (a letter that represents a number, such as x or b), a number (or constant such as 43 or 8), or a combination of both (such as $5y$, $3x^3$, or b^5). The number before a variable is called the *coefficient*. In the term $2y^3z^2$, **2** is the coefficient. Equations are *true* ($7 \times 9 = 63$), *false* ($5 - 3 = 7$), or *open* and called *open sentences* ($2x + 7 = 15$) because the equation contains a variable. The answers that make an open sentence true are contained in the *solution set*. The solution set for $2x + 7 \geq 15$, where x is an integer, is {4, 5, 6, ...} because substituting any of these numbers for x in this equation results in a true statement.

SOLVING EQUATIONS

When solving open sentences, remember to keep the *like terms* together. Terms are considered *like* when they have the same variable with the same exponent. The terms $7x$ and $4x$ are like terms. The terms $3x^2$ and $9x$ are not. Like terms can be simplified. The coefficients are combined but the variable remains the same. For example, $18x^2 + 3x^2 - 27x^2 = -6x^2$ because $18 + 3 - 27 = -6$. The variable that makes these terms like (x^2) is added on to the -6 in this example.

It is customary to write $1x$ as x. The exponent of 1 is never written. If you see a variable without an exponent, it is understood that the exponent is 1 and that the factor is used once.

When solving equations, get the terms with variables on one side of the equation and the constants on the other side. The equal sign separates the equation into two sides. You must perform the same action to both sides of the equation. For example, if you add $9y^4$ to $-9y^4$ on the left, you have to add $9y^4$ to the right side of the equation. When you have one variable term and one constant on opposite sides of the equation, then you can solve for x by using the multiplicative inverse of the coefficient on the variable. For example, if you have $7x = 42$, multiply both sides of the equation by $\frac{1}{7}$ and get $x = 6$. You can also divide both sides by 7 and get the same answer. Remember that dividing by 7 gives the same result as multiplying by $\frac{1}{7}$.

Example

A. Solve the equation $26 = 3x + 11$.

Get the term with the variable ($3x$) alone on the right, so add the inverse of 11 to both sides of the equation:

$$(-11) + 26 = 3x + 11 \; (-11)$$
$$15 = 3x$$

You want to know the value of x, so divide both sides of the equation by 3. Remember that dividing by 3 gives the same result as multiplying by one third:

$$\frac{15}{3} = \frac{3x}{3}$$

Perform the division and get $5 = x$.

It is always good to check your answer by substituting your answer for the variable to get a true statement. Check to see if $x = 5$ is correct in the preceding example.

$$26 = 3x + 11$$
$$26 \overset{?}{=} 3(5) + 11$$
$$26 \overset{?}{=} 15 + 11$$
$$26 \overset{\checkmark}{=} 26$$

Examples

B. Solve and check your answer: $5x + 24 = 8x - 66$

Add $-5x$ to both sides.

$$-5x + 5x + 24 = 8x - 66 - 5x$$
$$24 = 3x - 66$$

Add 66 to both sides.

$$66 + 24 = 3x - 66 + 66$$

Divide both sides by 3.

$$90 = 3x$$
$$30 = x$$

Check.

$$5x + 24 = 8x - 66$$
$$5(30) + 24 \stackrel{?}{=} 8(30) - 66$$
$$150 + 24 \stackrel{?}{=} 240 - 66$$
$$174 \stackrel{\checkmark}{=} 174$$

C. Solve for k.

$$\frac{k}{5} + 8 = 43$$

Add −8 to both sides.

$$\frac{k}{5} + 8 + (-8) = 43 + (-8)$$

$$\frac{k}{5} = 35$$

Now multiply by the multiplicative inverse of 1/5.

$$\frac{k}{5}\left(\frac{5}{1}\right) = 35(5)$$

$$k = 175$$

Now check your answer.

$$\frac{k}{5} + 8 = 43$$

$$\frac{175}{5} + 8 \stackrel{?}{=} 43$$

$$35 + 8 \stackrel{?}{=} 43$$

$$43 \stackrel{\checkmark}{=} 43$$

WRITING EQUATIONS FROM PROBLEMS

You may have to represent a value using variables given certain facts.

Examples

A. Over a period of three days during January it snowed 6 inches. If g represents the amount of snow already on the ground and t is the total amount of snow after the three days, write an equation showing the amount of snow on the ground after the new snowfall.

The equation can be $g + 6 = t$, or $t = g + 6$, or else $t - g = 6$. These equations are equivalent to each other.

B. The eighth grade class sold ice cream cones as a fund raiser for its class trip. The total income was $106.55 by selling large cones for $1.25 and medium cones for $.95. They made $22.80 on the medium cone sales. Write an equation and then solve it to find the number of large cones sold.

$$\$106.55 = \$22.80 + \$1.25l$$

$$\underline{-22.80 = -22.80}$$

$$83.75 = 1.25l$$

$$\frac{8375}{125} = \frac{125l}{125}$$

$$67 = l$$

Let l represent the amount of large cones sold at $1.25 each. Subtract 22.80 from each side. Divide both sides by 125 to get 67 as the number of large cones sold. The equation $83.75 = 1.25l$ became $8375 = 125l$ by multiplying both sides of the previous equation by 100. The same action of adding, subtracting, multiplying, or dividing terms must be done to both sides of an equation. Dividing by 125 in this example is equivalent to multiplying by $\frac{1}{125}$.

PROBLEM SOLVING

Here are some techniques you have used before to solve problems.

1. Read the problem carefully and find out what the problem is asking of you.
2. Use drawings, graphs, or charts.
3. Break the big problem into smaller ones.
4. Have you seen a problem like this before?
5. Work backward.
6. Guess and check.
7. Write an equation and then solve it.

To be a better problem solver, you need to work at solving more problems. Practice does make perfect. The more problems you solve, the better you get at it.

Examples

A. Kim's grandma was sick, so she went to visit and make her grandma some lunch. Grandma asked Kim to cook one 15-minute hard-boiled egg for her lunch. In her kitchen, Grandma's clock was not working, and Kim forgot to wear her watch. Grandma had two egg timers. They looked like miniature hourglasses. One had "7-minutes" printed on it, and the other had "11-minutes" printed on it. Using both of these timers, explain how to boil an egg for exactly 15 minutes.

Kim placed the egg in a pot of water on the stove. She turned over both timers at the same time. When the 7-minute timer ran out of sand, she turned on the burner because she knew that there were 4 minutes left in the 11-minute timer. When the 11-minute timer was empty, Kim turned it over and let the sand run out. The 4 minutes plus 11 minutes added to 15 minutes. So, Grandma got her 15-minute hard-boiled egg.

B. Three boxes are labeled "Apples," "Oranges," and "Apples and Oranges." Each label is incorrect. You may select *only one fruit from one box*. (No feeling or peeking permitted!) How can you label each box correctly?

First, decide which box will be the best from which to choose one fruit. For example, if you choose an orange from the box labeled "Apples," you don't know if the fruit actually in the box are oranges or apples and oranges. You should pick from the Apples and Oranges box first. The mislabeled boxes follow.

If you pick an apple, then the oranges are in the box labeled "Apples," and the mixed fruit is in the box labeled "Oranges."

If you choose an apple from the Apples and Oranges box:

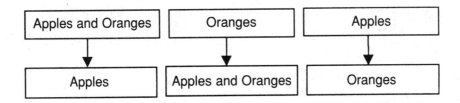

If you choose an orange from the Apples and Oranges box:

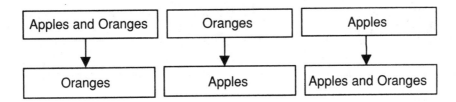

Remember that *all* the boxes are mislabeled, so each new label must be different from the old label.

C. A group of teenagers discovers that the product of their ages is 56,160. Find the ages of the teenagers.

At first it appears that there is not enough information to solve this problem. A teenager's age has "teen" in the number. The ages to consider are 13 through 19. One approach is *guess and check*. That could take a while unless you get lucky. Find the prime factors of 56,160. You will get $2 \times 2 \times 2 \times 2 \times 2 \times 5 \times 3 \times 3 \times 3 \times 13$. Now, multiply these prime factors together so they are all used and form numbers 13 through 19. The ages of the teenagers are 13 (a prime factor), 15 (3×5), 16 ($2 \times 2 \times 2 \times 2$), and 18 ($2 \times 3 \times 3$).

D. You are given a $10 \times 10 \times 10$ cube composed of $1 \times 1 \times 1$ minicubes glued together. If the outermost layer falls off, how many minicubes would have fallen off? Explain your answer.

Here is one way to solve this problem. The original cube is made of 1000 minicubes. If the outer layer falls off, minicubes from all six sides fall off. What is left is a smaller $8 \times 8 \times 8$ cube. This cube consists of 512 minicubes. The number of minicubes that fell off is $1000 - 512$ or 488.

You can also picture in your mind that the front and back minicubes fall off (2 sides $\times$ 10 cubes $\times$ 10 cubes or 200 minicubes). Then minicubes fall off from the left and right sides ($2 \times 10 \times 8$ or 160 minicubes). Finally, minicubes from the top and bottom fall off ($2 \times 8 \times 8$ or 128 minicubes). The total number that fell off is $200 + 160 + 128$ or 488 minicubes.

TEST YOUR SKILLS

1. Last weekend you made $33 babysitting. Your rates are $5 for the first hour and $3 for every hour after that. Three families relied on your services. How many hours did you work at babysitting?

2. Hypatia* and Ravi picked enchanted apples in the giant's garden. On their way home, they came to a troll who asked for one fourth of their apples plus 3 more. A little further down the road they came across a second troll who asked for one third of their apples plus 2 more. Again the road was blocked by yet a third troll who asked for half of their apples plus 1. Hypatia and Ravi came home with 2 apples. How many apples did they pick from the garden?

3. Solve the equation for k:

$$12k + 36 - 4k = 57 + 2k + 15$$

4. Simplify this expression:

$$4y^3 - 18 + 3y - 18y^3 + 11y$$

5. Solve for w and then check your answer:

$$\frac{5w}{4} + 2 = 72$$

6. The pet store is having a going-out-of-business sale, and you want to purchase a few new fish for your aquarium. You have a $5 limit to spend on lyre-tail mollies and fancy-tail guppies. The guppies are $.40 each, and the mollies are $.50 each. You want five guppies and want to buy mollies with the remaining money. Write an equation and then solve it to find the number of mollies you are able to purchase.

---

*Hypatia was a female mathematician who lived in Ancient Greece.

7. Imagine yourself as a space explorer and visitor to the planet Kipperdyng. The innkeeper rents you a room for seven days as you requested. Payment to the innkeeper will be made with the seven gold rings you have linked together in a chain, . What is the smallest number of rings you should cut open to pay one ring per day? The innkeeper cannot be trusted, so you cannot pay in advance. He will trade rings with you to make "change." Explain how the payment will be made daily.

8. Write an equation, then solve this problem. The small town of Pineville has a population of 2100. Each year the population increases by 60. Find the number of years it will take until the population reaches 3000.

9. Write an equation to solve this problem and then check your answer. If 3 times a number is increased by 12, the result is equal to 7 times that number.

10. The length of a rectangle is twice the width. If the length is increased by 4 and the width is diminished by 1, the new perimeter is 198. Find the dimensions of the original rectangle.

TEST YOUR SKILLS SOLUTIONS

1. Write an equation, then solve it.

   $$\$33 = \$5(3) + \$3x \qquad \text{x represents the number of hours at \$3.}$$
   $$33 = 15 + 3x$$
   $$\underline{-15 = -15}$$
   $$18 = 3x$$
   $$x = 6$$

   Subtract 15 from both sides.
   Divide both sides by 3.

   You worked 6 hours at $3 and 3 hours at $5. The total number of hours you worked was 9.

2. The best way to solve this problem is by working backward. If the third troll asked for half the apples plus 1 and they came home with 2 apples, then Hypatia and Ravi had six apples when they approached the third troll ($2 + 1 = 3$, and 3 doubled is 6).

   The second troll asked for one third plus 2. They had 6 to start + 2 extra = 8 apples. Eight represents two thirds of the number we are looking for:

   $$\frac{2}{3}x = 8$$

   $$2x = 24$$

   $$x = 12$$

   The number of apples they had before reaching the second troll was 12.

   The first troll asked for one fourth plus 3 apples. $12 + 3 = 15$. This number represents three fourths of the number originally picked.

   $$\frac{3}{4}x = 15$$

   $$3x = 60$$

   $$x = 20$$

   Hypatia and Ravi picked 20 apples from the garden.

3. $12k + 36 - 4k = 57 + 2k + 15$

$36 + 8k = 72 + 2k$ Combine like terms.

$\underline{-36 = -36}$ Subtract 36 from both sides.

$8k = 36 + 2k$

$\underline{-2k = -2k}$ Subtract $2k$ from both sides.

$6k = 36$ Divide both sides by 6.

$k = 6$

4. Combine the like terms when simplifying an expression. This is *not* an equation, so you cannot solve for the variable.

$$4y^3 - 18 + 3y - 18y^3 + 11y$$

Combine like terms $4y^3$ and $-18y^3$ by adding to get $-14y^3$. Combine $3y$ and $11y$ to get $14y$. You cannot combine these terms because the exponents on the y variables are not the same. The answer is $-14y^3 + 14y - 18$.

5. Here is one way to solve for w: Checking:

$\dfrac{5w}{4} + 2 = 72$ $\dfrac{5(56)}{4} + 2 \overset{?}{=} 72$

$\dfrac{5w}{4} = 70$ $\dfrac{280}{4} + 2 \overset{?}{=} 72$

$5w = 280$ $70 + 2 \overset{?}{=} 72$

$w = 56$ $72 \overset{\checkmark}{=} 72$

6. Let m = the number of mollies.

$$0.40(5) + 0.50m = 5$$
$$2 + 0.5m = 5$$
$$0.5m = 3$$
$$m = 6$$

You spend \$2 on five guppies and \$3 on six mollies.

7. If you break the third link, there are three sections to be used for payment.

   On the first day, pay the innkeeper the opened link. On day two, pay with the two-link section and get the one link back. On day three, give the one link to the innkeeper so that he has three links. On day four, get back all the links and give him the four-link section. On day five, give him the one link. On day six, get back the one link and give him the two-link section. On day seven, give him the one link. The most links you need to cut open is one.

8. Let y represent the number of years before the population reaches 3000.

$$2100 + 60y = 3000$$
$$60y = 900$$
$$y = 15$$

Check:

$$2100 + 60(15) \stackrel{?}{=} 3000$$
$$2100 + 900 \stackrel{?}{=} 3000$$
$$3000 \stackrel{\checkmark}{=} 3000$$

9. Let x be the number.   Check:

$$3x + 12 = 7x$$ $$3x + 12 = 7x$$
$$12 = 4x$$ $$3(3) + 12 \stackrel{?}{=} 7(3)$$
$$3 = x$$ $$9 + 12 \stackrel{?}{=} 21$$
                        $$21 \stackrel{\checkmark}{=} 21$$

10. Draw two rectangles and label them "old" and "new."

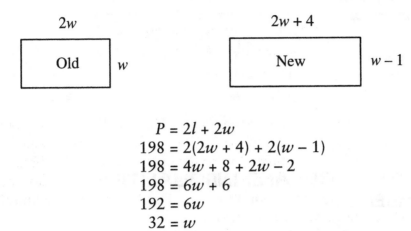

$$P = 2l + 2w$$
$$198 = 2(2w + 4) + 2(w - 1)$$
$$198 = 4w + 8 + 2w - 2$$
$$198 = 6w + 6$$
$$192 = 6w$$
$$32 = w$$

The original rectangle has dimensions of 32 for the width and 64 for the length.

Chapter 10

Graphing

GRAPHING EQUALITIES AND INEQUALITIES ON THE NUMBER LINE

The symbols < (less than), > (greater than), ≤ (less than or equal to), and ≥ (greater than or equal to) denote *inequality*. The solution sets of inequalities and equations can be graphed on the real number line.

Examples

A. Graph the solution to $3x - 7 \geq 26$.

The solution is $x \geq 11$. On the number line, blacken the point that represents 11 as well as an arrow that shows that the numbers greater than 11 are also included in the solution set.

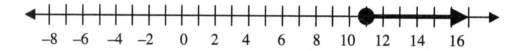

B. Graph the solution to $3x - 7 < 26$.

This time, solving the equation gives $x < 11$. The solution set to this inequality contains all the numbers less than but not equal to 11.

From these examples you see that if a number makes the open sentence true, the point is blackened. If not, an open circle represents the number and a shaded arrow is drawn in the direction of the other numbers in the solution set.

C. Graph the solution to $4x + 11 \geq 13x - 7$.

Inequalities are solved in the same way as equations.

Add $-4x$ to both sides:	$(-4x) + 4x + 11 \geq 13x - 7 + (-4x)$
Add 7 to both sides:	$(+ 7) + 11 \geq 9x - 7 (+ 7)$
Divide both sides by 9:	$18 \geq 9x$
Solution:	$2 \geq x \ \ or \ \ x \leq 2$

This graph shows that all the numbers less than or equal to 2 will make this inequality a true statement when substituted for x.

GRAPHING ON THE COORDINATE SYSTEM

Two number lines placed perpendicular to each other form a Cartesian graph, named after René Descartes. He was a famous French mathematician and philosopher. The horizontal number line is now called the *x-axis,* and the vertical number line is called the *y-axis.* These *axes* break the plane into four quadrants. The diagram shows the signs of the coordinates in each of the quadrants. The first number is the *x-coordinate* and the second is the *y-coordinate.* This pair is grouped by parentheses. *Order matters* in the pair. The point (5, 2) is *not* the same point as (2, 5). The two axes intersect at the *origin,* whose coordinates are (0, 0). The *x*-axis heads right toward +∞ and left toward −∞. The numbering on the *y*-axis is similar to a thermometer where upward is +∞ and downward is −∞.

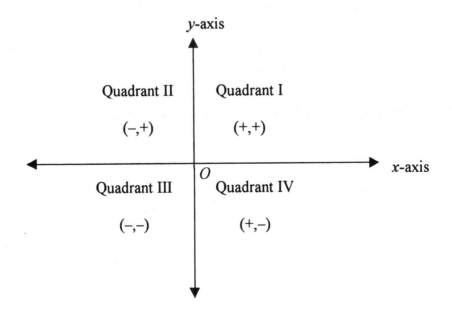

PLOTTING POINTS

To plot points on this graphing system, find the location of the first number on the (horizontal) x-axis. From that number, go either up or down to find the location of the second number on the (vertical) y-axis.

Example

A. Plot and label the following points on the graph.

$$A\ (4, 2),\quad B\ (-3, 1),\quad C\ (0, -6),\quad D\ (-5, -7),\quad E\ (7, 0),\quad F\ (2, -4)$$

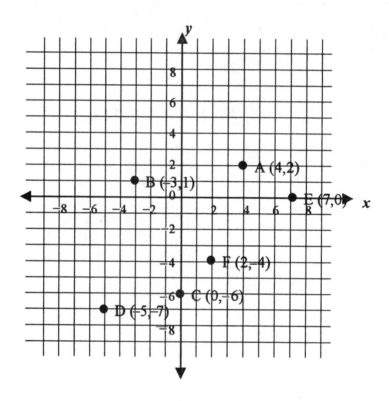

SLOPE OF LINES

The *slope* tells you how much pitch there is to the line. Lines have equations of the form $y = mx + b$. The variables x and y are coordinates from one of the points on the line, m is the slope of the line, and b is the y-intercept or the point on the line that intersects with the y-axis.

Look at the line from left to right. A positive slope increases (⟋) and a negative slope decreases (⟍). Horizontal lines have no slope, and vertical lines have a slope that is undefined. When you are finding the slope of a line, you are looking at the vertical change over the horizontal change in the line. The slope of a line is determined by the ratio of the $\frac{rise}{run}$ if the slope is positive or $\frac{fall}{run}$ if the slope is negative. The *run* is counted toward the *right* in this chapter.

Examples

A. Find the slope of the line on the graph. The scale of the graph is one unit.

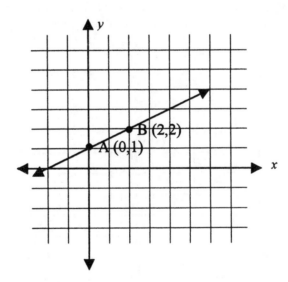

Locate two points on the line. To get from A to B, go up one unit so the numerator of the ratio is 1. To get from A to B, move 2 units to the right. The rise is 1 and the run is 2. The slope of this line is $\frac{1}{2}$. Because the line moves up as you follow it from the left to the right, the slope is positive.

B. On the graph, draw a line that has a slope of –4. The scale is one unit.

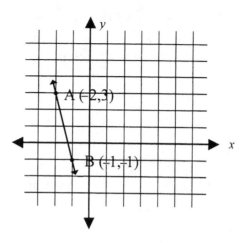

–4 can be rewritten as $\frac{-4}{1}$. Locate an arbitrary point on the graph. We will call this point A. From A, count *down* 4 units because the slope is negative, and then move one unit to the right. This is your second point of the line; we will call it point B. Construct the line with both points.

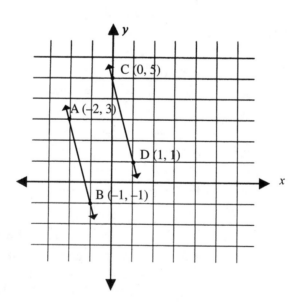

The last example provided only one of many possible answers. Depending on the first point you choose, you can construct an infinite number of lines that all have the same slope. Suppose you chose your first point, C, at (0, 5). Your second point, D, has coordinates of (1, 1). These two lines are parallel. Parallel lines have the same slope.

WRITING EQUATIONS OF LINES

Vertical and horizontal lines have special equations. In the graph (scale of one unit), the vertical line has the equation $x = 5$ because every x-coordinate of every point that lies on this line is 5. The equations of horizontal lines are similar in that every point on the line $y = -2$ has a y-coordinate of -2.

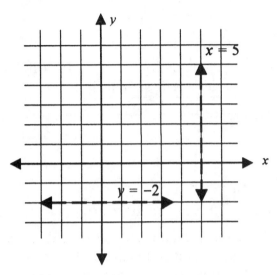

Examples

A. Lines m and k are perpendicular. Find the slope of each line, and describe how the slopes are related.

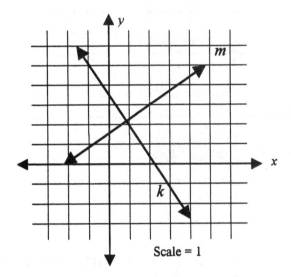

The slope of line m is $\frac{+2}{3}$, and the slope of line k is $\frac{-3}{2}$. The slopes are negative reciprocals of each other. Their product is -1.

B. Write the equation of this line.

Use the form $y = mx + b$. The coordinates of the points are $(0, 2)$ and $(5, -1)$. Counting from the first point to the second, get the slope of $\frac{-3}{5}$. The y-intercept is 2. So, the equation of the line is

$$y = \frac{-3}{5} x + 2$$

The variables x and y remain in the equation. If you pick a point on that line and substitute the coordinates, you should get a true statement.

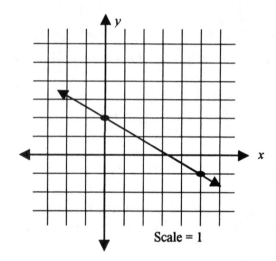

Scale = 1

You can use the equation of a line to check if a point lies on that line.

Examples

C. Does the point $(3, -4)$ lie on the line $y = 2x - 5$?

Substitute the coordinates in the equation.

$$y = 2x - 5$$
$$-4 \overset{?}{=} 2(3) - 5$$
$$-4 \overset{?}{=} 6 - 5$$
$$-4 \neq 1$$

Therefore, $(3, -4)$ does not lie on the line.

D. Write the equation of the line that passes through the points (–3, 5) and (0, 1).

This is possible without using a graph. The slope is found by finding

$$\frac{\text{Difference in the } y\text{-coordinates}}{\text{Difference in the } x\text{-coordinates}} = \frac{y_1 - y_2}{x_1 - x_2}$$

Choose (–3, 5) to be your first point and (0, 1) to be your second point. The point order matters here! Notice the subscripts.

$$\frac{y_1 - y_2}{x_1 - x_2} = \frac{5 - 1}{-3 - 0} = \frac{4}{-3} = \frac{-4}{3}$$

The slope is $\frac{-4}{3}$. The second point states the y-intercept as +1. If a point has an x-coordinate of zero, then the point lies on the y-axis and is the y-intercept. So, the equation of the line is

$$y = \frac{-4}{3}x + 1$$

E. List three points that lie on the line $8x + 2y = 10$.

Make a table of values, then choose three values for x. It is a good idea to choose a positive x, a negative x, and 0. Then find the y-value by solving the equation.

x-Value	y-Value	Point
–1	$8(-1) + 2y = 10$ $-8 + 2y = 10$ $2y = 18$ $y = 9$	(–1, 9)
0	$0 + 2y = 10$ $y = 5$	(0, 5)
1	$8 + 2y = 10$ $2y = 2$	(1, 1)

Therefore, the points (–1, 9), (0, 5), and (1, 1) lie on the given line.

F. Graph the line $y = x$.

Each point that lies on this line has the same value for the x- and y-coordinates. Some points on this line are $(-2, -2)$, $(0, 0)$, $(4, 4)$, etc. Another way to look at this problem is to notice that the coefficient on x is 1, so the slope is 1. There is no y-intercept, so that is 0, or the origin $(0, 0)$.

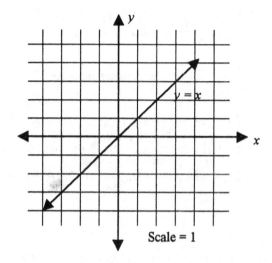

TEST YOUR SKILLS

1. Write the equation of the line that is parallel to the line $y = \frac{-3}{2}x + 8$ and passes through the point $(1, 1)$.

2. Write the equation of the line of the hypotenuse of a right triangle whose area is 36 square units. The legs of this triangle are segments of the x- and y-axes. The right angle is at the origin. Another vertex is at $(6, 0)$.

3. Which slope is steeper: $\frac{-3}{5}$ or -2. Explain your answer.

4. You are building a small shed to house two garbage cans. The side of the shed looks like a trapezoid. The pitch of the roof is $\frac{1}{3}$ to keep off the snow and rain. The back wall will be 4 feet tall and the shed will be 3 feet deep. How tall will the front door be?

5. The area of a parallelogram is 6 square units. Three of the vertices are $(1, -2)$, $(1, -4)$, and $(4, -1)$. What are the coordinates of the fourth vertex?

6. Write an equation for the line that is perpendicular to $y = \frac{-1}{2}x + 8$ and passes through the point $(3, 3)$.

7. Using a table of values, find five points on the line $y = 3x - 2$.

8. Find the y-intercept of the line $15x - 3y = 6$.

9. Fill in the missing coordinate values for the line with the equation

$$y = 5x - 2$$

x	y
	–2
2	
	–7
4	
	–17

10. One summer, Matt and Jesse read the book *Treasure Island* and decided to create their own map. The boys mapped Matt's backyard, including the location of landmarks, placed a few favored items in a box, and wrote down the equations of two lines instead of an **X** as the location of their treasure. The intersection of these lines was the location of the box. What are the coordinates of the treasure if the lines are $x + y = 6$ and $y = 2x$?

TEST YOUR SKILLS SOLUTIONS

1. The equation of a line is $y = mx + b$. The values of x and y come from the given point (1,1). The slope of the new line will have the same slope. Substitute the numbers for the variables and get

$$1 = \frac{-3}{2}(1) + b$$

Then solve the equation for b.

Multiply both sides of this equation by 2.

$$2 = -3 + 2b$$

Add 3 to both sides.

$$5 = 2b$$

Divide both sides by 2.

$$\frac{5}{2} = b$$

The equation of the line parallel to the given line is

$$y = \frac{-3}{2}x + \frac{5}{2}$$

You can also solve this another way by adding $\frac{3}{2}$ to both sides of the equation

$$\frac{2}{2} + \frac{3}{2} = b$$

$$\frac{5}{2} = b$$

The equation is

$$y = \frac{-3}{2}x + \frac{5}{2}$$

2. We know that the base of this triangle is 6. The only possible point for the third vertex is at (0, 12) to get an area of 36 square units when using the formula $A = \frac{bh}{2}$.

   Use the points (0, 12) and (6, 0) to get the slope of $\frac{-12}{6}$ or –2. The y-intercept is 12 (from the point (0, 12). That means the equation of the hypotenuse is $y = -2x + 12$.

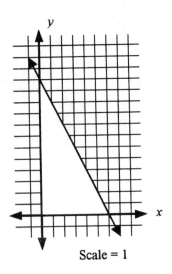

Scale = 1

3. Rewrite both slopes in rational form with a common denominator.

$$\frac{-3}{5} \text{ and } \frac{-10}{5}$$

   The steeper slope will be –2 because over the same distance of 5 units, the slope of –2 drops 10 units compared to a 3-unit drop.

4. Draw a diagram of the shed. The height of the back wall is 4 units high (0, 4). From this point (0, 4), count up 1 unit then right 3 units. The front door of the shed will be 5 feet tall.

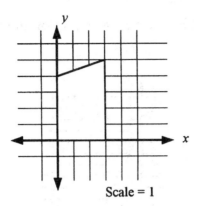

Scale = 1

5. The only possible location for the fourth vertex is (4, −3). It creates a parallelogram.

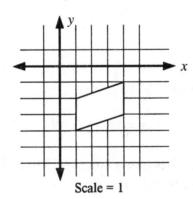

Scale = 1

6. The line perpendicular to the given line has a slope of 2. This slope is the negative reciprocal of $\frac{-1}{2}$. Substitute the slope and the x- and y-value from the point (3, 3) into the equation $y = mx + b$, then solve for b.

$$y = mx + b$$
$$3 = 2(3) + b$$
$$3 = 6 + b$$
$$-3 = b$$

The equation is $y = 2x - 3$.

7. Make a table and choose five values for x.

x-Value	y-Value	Point
–2	$3(-2) - 2 = -8$	$(-2, -8)$
–1	$3(-1) - 2 = -5$	$(-1, -5)$
0	$3(0) - 2 = -2$	$(0, -2)$
1	$3(1) - 2 = 1$	$(1, 1)$
2	$3(2) - 2 = 4$	$(1, 4)$

8. Rewrite the equation in the form $y = mx + b$.

$$15x - 3y = 6$$
$$-3y = -15x + 6$$
$$y = 5x - 2$$
$$y = 5x + -2$$

The y-intercept is –2.

9. In the equation $y = 5x - 2$, substitute the value given in the chart, then solve the equation for the missing variable.

x	y	Work
0	–2	$-2 = 5x - 2$ $0 = 5x$ $x = 0$
2	8	$y = 5(2) - 2$ $y = 10 - 2$ $y = 8$
–1	–7	$-7 = 5x - 2$ $-5 = 5x$ $x = -1$
4	18	$y = 5(4) - 2$ $y = 20 - 2$ $y = 18$
–3	–17	$-17 = 5x - 2$ $-15 = 5x$ $x = -3$

10. On a graph, construct the lines. Then, find the intersection point. The treasure is buried at the intersection point (2, 4).

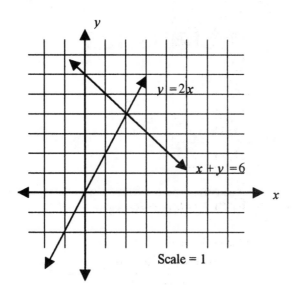

You can also solve this algebraically by substituting the equation $y = 2x$ into the equation $x + y = 6$ to get $x + 2x = 6$ and then solving to get $x = 2$. Substitute 2 for x in either one of the equations and get $y = 4$.

$$x + y = 6$$
$$2 + y = 6 \qquad \text{or} \qquad y = 2x$$
$$y = 4 \qquad\qquad\qquad y = 2(2)$$
$$\qquad\qquad\qquad\qquad\qquad y = 4$$

Chapter 11

Probability and Statistics

PROBABILITY

Because we cannot predict an occurrence of a particular event with complete certainty, we use *probability*. The probability of an event is expressed as a number between 0 and 1, where E is the event:

$$P(E) = \frac{\text{\# of favorable outcomes}}{\text{Total \# of outcomes}}$$

Probabilities can be written in any form such as fractions, decimals, or percents. If an event will never happen (pigs sing opera), then the probability is 0. If the event will definitely happen (it snows in winter in the northern hemisphere), then the probability is 1.

The results of a coin toss have a *sample space* that is {H, T}. The *relative frequency* of getting heads is 50%. This does not mean that if you tossed a coin 10 times that you would always get 5 heads and 5 tails. By performing this experiment and keeping a tally of the results, you can see that the *cumulative frequency* shows that you will get a head approximately half of the time and a tail approximately half of the time.

Example

A. Suppose that you roll a *fair* (each number is equally likely to appear), six-sided die once. The *outcomes* can be shown as the set {1, 2, 3, 4, 5, 6}. Find the following:

(a) P (4), (b) P (even number), (c) P (prime number), (d) P (even number), and (e) P (a number < 5).

(a) P (4) = $\frac{1}{6}$ because there is only one 4 on the die of six numbers that could appear.

(b) P (even number) = P (2) or P (4) or P (6) = $\frac{1}{6} + \frac{1}{6} + \frac{1}{6} = \frac{3}{6} = \frac{1}{2}$. These events (getting a 2, 4, or 6) are *mutually exclusive* because they cannot happen at the same time. You are rolling the die *once*, but three possible numbers could appear. Connecting the probabilities of the different numbers with *or* means you add the probabilities.

(c) P (prime number) = P (2) + P (3) + P (5) = $\frac{1}{6}$ + $\frac{1}{6}$ + $\frac{1}{6}$ = $\frac{3}{6}$ = $\frac{1}{2}$.

(d) P (multiple of 5) = P (5) = $\frac{1}{6}$.

(e) P (a number < 5) = P (1) + P (2) + P (3) + P (4) = $\frac{1}{6}$ + $\frac{1}{6}$ + $\frac{1}{6}$ + $\frac{1}{6}$ = $\frac{4}{6}$ = $\frac{2}{3}$.

The answer in (d) can be solved in another way using the *complement rule*. The sum of the probabilities of the individual events is 1. So,

$$1 = P \text{ (not getting 5)} + P \text{ (getting 5)}$$

$$1 - P \text{ (not getting 5)} = P \text{ (getting 5)}$$

$$1 - \frac{5}{6} = \frac{1}{6}$$

Examples

B. Find P ($x \geq 5$).

$$P (x \geq 5) = 1 - P (x < 5) = 1 - \frac{2}{3} = \frac{1}{3}$$

Or else, find

$$P (5) + P (6) = \frac{1}{6} + \frac{1}{6} = \frac{2}{6} = \frac{1}{3}$$

C. A box contains four black and three white marbles. You are to draw one marble from the box. What are the probabilities for each color?

A tree diagram can help. P (w) = $\frac{3}{7}$ and P (b) = $\frac{4}{7}$.

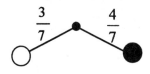

EXPERIMENTS WITH REPLACEMENT

Examples

D. A box contains two yellow and three green balls. You will draw one ball and record its color. Replace the ball in the box and draw a second ball. Find the probability that you get a green ball on the first draw and a yellow on the second.

To find P (G, Y), draw a tree and label the probabilities.

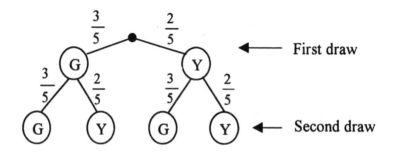

In this case, we want *both* events to occur. This means we want P (G) *and* P (Y). So, follow the branches of the tree and multiply the probabilities.

$$P \text{ (G, Y)} = \frac{3}{5} \times \frac{2}{5} = \frac{6}{25}$$

E. Using the tree in the preceding problem, find the probability that both balls are the same color.

The outcomes we are looking for are P (G, G) or P (Y, Y). Remember to add when you see *or* and multiply when you see *and*.

$$P \text{ (both balls same color)} = P \text{ (G, G)} + P \text{ (Y, Y)}$$

$$= \left(\frac{3}{5} \times \frac{3}{5}\right) + \left(\frac{2}{5} \times \frac{2}{5}\right)$$

$$= \frac{9}{25} + \frac{4}{25} = \frac{13}{25}$$

EXPERIMENTS WITHOUT REPLACEMENT

Example

F. A bag contains five blue and seven white marbles. Draw three marbles from the bag. Without replacing the marbles after each draw, find the probability that: (a) all three marbles are the same color, (b) the last marble is blue, and (c) the first and last marbles are the same color.

First, draw the tree. When calculating the probabilities remember that one less marble is in the bag after each draw. Multiply the branches of the tree to get the probability of that occurrence.

(a) The probability that all three marbles are the same color is the same as

$$P(B, B, B) + P(W, W, W) = \frac{1}{22} + \frac{7}{44} = \frac{2 + 7}{44} = \frac{9}{44}$$

(b) $P(x, x, B) = P(B, B, B) + P(B, W, B) + P(W, B, B) + P(W, W, B)$

$$= \frac{1}{22} + \frac{7}{66} + \frac{7}{66} + \frac{7}{44} = \frac{30 + 70 + 70 + 105}{660} = \frac{275}{660} = \frac{5}{12}$$

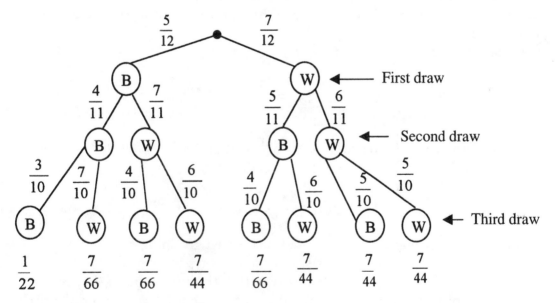

(c) P (first and last same color) $= P(B, B, B) + P(B, W, B) + P(W, B, W) + P(W, W, W)$

$$= \frac{1}{22} + \frac{7}{66} + \frac{7}{44} + \frac{7}{44} = \frac{30 + 70 + 105 + 105}{660} = \frac{310}{660} = \frac{31}{66}$$

Always reduce your fractions to lowest terms for your final answer.

COMBINATIONS AND PERMUTATIONS

There are five books on a shelf, each a different color—red, blue, green, yellow, and white. How many different ways can you arrange these five books on the shelf? There are five choices for the first position, four books for the second, three for the third, two for the fourth, and one for the fifth. There are $5 \times 4 \times 3 \times 2 \times 1 = 120$ or $5!$ (read as *five factorial*) different ways to arrange the books.

Now suppose that, of these five books, you want to choose three to put on your desk. How many ways can you choose three books? Use the *combination formula* $_nC_x$ where n is the number of items from which you are choosing x items. Your scientific calculator can perform combinations. To calculate $_5C_3$, substitute 5 for n and 3 for x.

$$_nC_x = \binom{n}{x} = \frac{n!}{x!(n-x)!}$$

$$_5C_3 = \binom{5}{3} = \frac{5!}{3!\,2!} = \frac{5 \times 4 \times 3 \times 2 \times 1}{3 \times 2 \times 1 \times 2 \times 1} = 10$$

There are ten ways to choose three of the five books. This is a *combination* where the order does *not* matter. So, the arrangements *red-blue-green, green-blue-red, blue-green-red, red-green-blue, green-red-blue,* and *blue-red-green* are considered the same and counted as one way.

If the order does matter, use the *permutation formula*

$$_nP_x = \frac{n!}{(n-x)!}$$

Now, how many ways are there to choose three of the five books?

$$_5P_3 = \frac{5!}{2!} = \frac{5 \times 4 \times 3 \times 2!}{2!} = 60$$

The six different arrangements of the red-blue-green grouping are now considered different from each other and counted as six of the 60 ways to arrange those three books. An easy way to compute a permutation is to start with the number n (in this case 5) as the first of a list of factors, each subsequent factor is one less, to be multiplied. The number of factors is x (in this case 3).

$$_5P_3 = 5 \times 4 \times 3 = 60$$

Examples

A. In a horse race, the first three horses to cross the finish line are in the *win, place,* or *show* positions. If ten horses are racing, how many possible winning orders are there?

Ten horses could win. That leaves nine horses to place and eight horses to show. Calculate the permutation $_{10}P_3 = 10 \times 9 \times 8 = 720$ possible orders.

B. Of the eight players on a basketball team, how many ways can the coach choose five players for the court?

$$_8P_5 = 8 \times 7 \times 6 \times 5 \times 4 = 6720 \text{ ways}$$

C. Abbey, Bill, Cara, and Derek are going tobogganing. Cara does not want to sit in front. How many ways can they be arranged?

$$\underline{\quad 3 \quad} \bullet \underline{\quad 3 \quad} \bullet \underline{\quad 2 \quad} \bullet \underline{\quad 1 \quad} = 18$$

There are three choices to sit in the first seat because Cara does not want to sit there. That leaves three choices for the second seat, two choices for the third seat, and one choice for the last seat.

COUNTING PRINCIPLE

Example

A. The school lunch menu has the following choices. How many different lunches can be made with one choice from each column?

Sandwich	Drink	Dessert
Hamburger	Milk	Ice cream
Pizza	Soda	Cookies
Grilled cheese		Fruit

A tree can show the lunch choices.

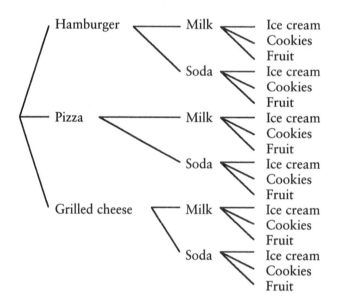

The *counting principle* states that if one activity can occur in m ways and another activity can occur in n ways, then both activities can occur in mn ways. In this example, there are $3 \times 2 \times 3$ or 18 different lunches.

B. What is the probability that a lunch consists of pizza, milk, and dessert?

Calculate

$$\frac{\text{Specific lunches}}{\text{Total number of lunches}} = \frac{1 \times 1 \times 3}{18} = \frac{3}{18} = \frac{1}{6}$$

C. Spin the spinner and toss a coin once. What is the probability you will not land on A and the coin will be tails?

Of the eight letters, three are A. That leaves five other letters. The probability of getting tails is one half.

$$P \text{ (not A, tails)} = \frac{5}{8} \times \frac{1}{2} = \frac{5}{16}$$

SAMPLE-BASED PROBABILITY

If you know the results of a sample population, you can make a prediction on a larger population.

Example

A. Twenty students were polled on their eye color. These are the results.

Blue	Brown	Green
12	6	2

Predict the number of students with green eyes out of 250 students.

Using a proportion is one way to solve this type of problem.

$$\frac{2}{20} = \frac{x}{250}$$

$$20x = 2(250)$$

$$20x = 500$$

$$x = 25$$

If 250 students were polled, 25 would probably have green eyes.

Another way to solve this problem is to multiply the ratio of

$$\frac{\text{\# with green eyes}}{\text{\# in the sample}} \times \text{\# in the larger population}$$

$$\frac{2}{20} \times 250 = 25$$

STATISTICS

The science of collecting, organizing, and reporting data is called *statistics*. Collecting data from samples of the population is important for the report to be fair. To find the average November heating bill in the United States, would you choose only those families who live in the northeastern United States where the winters are colder than elsewhere in the country?

MEAN, MEDIAN, MODE, AND RANGE

The *mean* is the average of a set of data. After data is organized in value from lowest to highest, the *median* is the middle number. The number appearing most frequently in a set of data is called the *mode*. In a set of data, the spread of the numbers given by the lowest and highest values is called the *range*.

Examples

A. Rushad's test scores are 89, 93, and 84. What must he get on the fourth test to get an average of 90? What is the highest average he could get for this quarter?

His four scores divided by 4 should equal 90. Another way of saying this is that his test score total should equal 360. So far the total of his scores is 266. He needs to get 94 on his fourth test to get a 90 average.
If he got 100 on his fourth test, his average would be 91.5.

B. Find the mode of the following running times of the track team. Is the mode much different from the mean time? What is the range of times?

1:15 2:07 1:21 1:40 1:15 1:20 1:35 1:21 1:30 1:21
1:56 1:17 1:32 1:58 1:32

The time 1:21 is the *mode*. When finding the average, change the times into minutes. Remember that time uses a base 60 system. The *mean* is 92 minutes or 1:32. In this instance, the mode is less than the average. The *median* of 1:30 is closer to the average than the mode. The range is 1:15 to 2:07.

1:15 1:15 1:17 1:20 1:21 1:21 1:21 1:30 1:32 1:32
1:35 1:40 1:56 1:58 2:07

GRAPHING DATA

Graphs are used to display data. Readers get a lot of information from a graph quickly. The type of graph used depends on the data. There are pie or circle graphs, bar graphs, line graphs, or pictographs. Double bar or double line graphs show a comparison. Graphs need a title, must have their axes labeled, need correct scales on the axes, and need a key for the reader.

Examples

A. Your brother's scout troop #812 is having a paper drive in competition with another scout troop. Over eight days, troop #812 collected the following amounts of newspaper in pounds:

Day 1	Day 2	Day 3	Day 4	Day 5	Day 6	Day 7	Day 8
120	100	85	90	110	80	100	95

The other troop #637 collected the following pounds of papers:

Day 1	Day 2	Day 3	Day 4	Day 5	Day 6	Day 7	Day 8
80	90	85	90	70	60	100	110

Make a double line graph depicting their collections.

Each line of the graph represents a troop. The horizontal axis should be labeled "Days" because you want to see a change in amount over the days, the independent variable. The vertical axis is labeled "Weight in Pounds," the variable that changes.

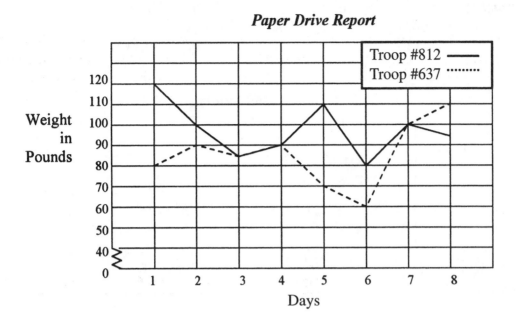

B. Create a double bar graph with the same data on the paper drive.

The axes of the double bar graph will be the same. Instead of lines, the bars comparing the amounts in pounds of the newspapers for each day are next to each other. There is a space between the bars for each day.

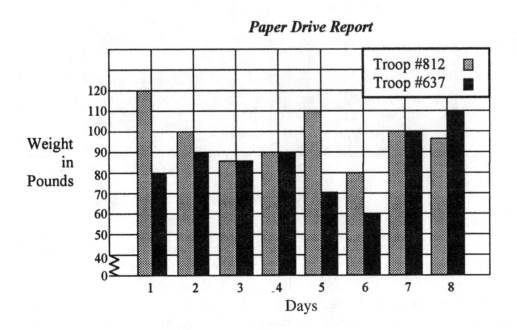

A *stem and leaf plot* of the scores of a test follows. To the left of the line is the tens digit and to the right of the line are the ones or units digit of the score. The score of 94 looks like 9|4. Using these data, we can create a histogram that will show the number of values in a particular range.

Test Scores of 60 Students

10	0, 0, 0, 0
9	9, 9, 6, 5, 4, 3, 3, 3, 2, 0
8	8, 8, 7, 6, 6, 6, 5, 3, 3, 3, 2, 2, 1, 1, 1, 1, 0
7	9, 8, 8, 8, 6, 6, 5, 5, 4, 3, 2, 1, 0
6	7, 6, 4, 3, 2, 2, 1, 0, 0
5	9, 7, 4, 3, 2, 1, 1

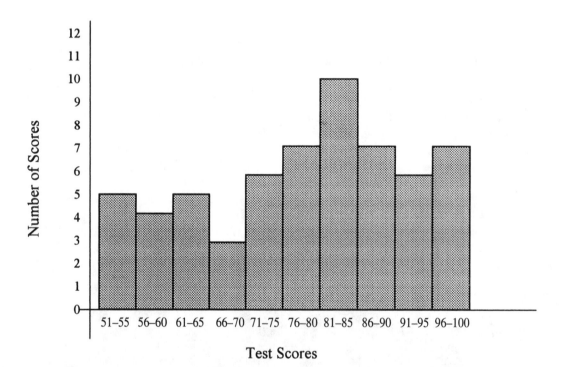

The data from the stem and leaf graph can be represented using a histogram, which is a vertical bar graph with the bars next to each other. Notice that the scores of the test for each bar have a range of five scores. Having a range rather than a single number representation on the horizontal axis is typical of a histogram. This graph shows that the largest number of students scored 81 to 85 and that seven students scored higher than 95 on this test.

TEST YOUR SKILLS

Answer questions 1 through 3 based on the following information:

A fair coin and a fair die are tossed at the same time.

1. Draw a tree showing all the possible results.

2. Find the probability of getting a head and an even number.

3. What is the probability of getting a tail and a number greater than 4?

4. On the table are two boxes. Box 1 contains three black and four white marbles. Box 2 contains two white and two black marbles. Reach into box 1 and get one marble. Without looking at it, drop it into box 2. What is the probability of getting a black marble from box 2?

5. Uncle Harry works as a clown at children's parties on the weekends. He has four clown shirts that are blue, purple, red, and yellow. He has three pairs of clown pants that are striped, plaid, and checked. What is the probability that he will select at random a yellow shirt and checked pants?

6. Six students in your music class play the violin. How many ways can your teacher choose two violinists to play a duet, where one student plays the melody and the other plays the harmony?

7. A new breakfast cereal, Krunchy Kritters, was taste-tested. The company took samples to the largest mall in 30 states. At each location, 20 people are asked for their opinion. Here are the results.

Age	Like	Dislike
0–15	163	26
16–30	90	37
31–45	81	43
46–60	86	74
Total	420	180

If 15,000 people were polled, how many 16–30 year olds in the United States would most likely enjoy this cereal?

8. As a community service project, the eighth grade class collected mittens and socks for neighborhood shelters. The students collected 72, 77, 84, 96, and 161 mittens from mall patrons on Tuesday through Saturday, respectively. On those same days, they collected 36, 28, 34, 57, and 85 pairs of socks. Draw a bar graph that displays this information. Label the graph appropriately.

9. The Ducky Pen Company makes a pen with a clear barrel. The Quality Control Department will find 30 out of 500 pens examined per day that do not have an ink cartridge inserted during assembly. If the company produces 20,000 pens in a 5-day period, how many pens will be defective?

10. The germination rate of a certain flower seed is 4:7; therefore, of seven seeds planted, four will produce a plant. How many seeds must be planted to ensure that there will be a minimum of 1500 plants available at the nursery?

TEST YOUR SKILLS SOLUTIONS

1. The tree of all the outcomes looks like this with 12 outcomes.

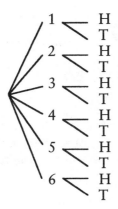

2. P (H, even #) $= \dfrac{1}{2} \times \dfrac{1}{2} = \dfrac{1}{4}$

There are 3 choices (H, 2), (H, 4), and (H, 6) from 12 outcomes.

3. P (T, $x > 4$) $= \dfrac{1}{2} \times \dfrac{1}{3} = \dfrac{1}{6}$

You could get (T, 5) or (T, 6) from 12 outcomes.

4. Draw a tree that shows the outcomes. In this problem, you are performing two steps. First you draw a marble from box 1. Your choice affects the contents of box 2.

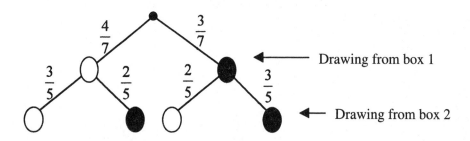

Multiply the branches of the tree that end in a black marble. Since there are two, add the products together to find the probability of getting a black marble from box 2.

$$\left(\frac{4}{7} \times \frac{2}{5}\right) + \left(\frac{3}{7} \times \frac{3}{5}\right) = \frac{8}{35} + \frac{9}{35} = \frac{17}{35}, \text{ which is about } 50\%$$

5. Uncle Harry has 4×3 or 12 different outfits. The probability he will pick a yellow shirt and checked pants is $\frac{1}{4} \times \frac{1}{3} = \frac{1}{12}$.

6. The two chosen students have different music to play. There are six students from which to choose to play the melody. This leaves five students who can play the harmony. So, there are 6×5 or 30 ways to pick two students.

7. Out of 600 people, 90 liked the cereal. If 15,000 people were tested, 2250 people in the 16- to 30-year-old range would like the Krunchy Kritters cereal.

$$\frac{90}{600} = \frac{x}{15,000}$$

$$x = 2250$$

8. This is one possible graph. You can see from this graph that the collections were better on Friday and Saturday.

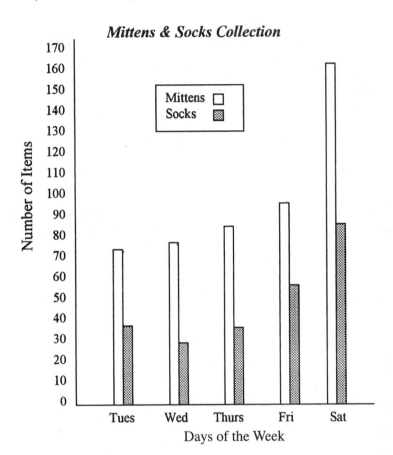

9. A proportion can be used to solve this problem.

$$\frac{30}{500} = \frac{3}{50} = \frac{x}{20{,}000}$$

$$5x = 6000$$

$$x = 1200$$

Quality Control can expect 1200 defective pens in 5 days.

10.

$$\frac{4}{7} = \frac{1500}{x}$$

$$4x = 10,500$$

$$x = 2625$$

The nursery should plant a minimum of 2625 seeds to get 1500 plants.

Chapter 12

Sample Test and Solutions

ABOUT THE TEST

The test will be administered over two days. Each day the time limit is approximately 70 minutes. On the first day, you have about 35 minutes for the multiple-choice questions and another 35 minutes for the 6 questions from Part II. On the second day, you have about 70 minutes to answer the other 12 questions. Calculators are not allowed on the multiple-choice questions; however, you will be able to use your calculator for all the questions in Part II. This includes those 6 questions on the first day.

Read all the questions carefully. First answer the questions that seem the easiest. Then go back and answer the others. In that way, you will not be wasting time. Read the choices of the multiple-choice questions carefully and disregard those that do not make sense to narrow down the possible correct answers. You have a better chance of choosing the correct answer from two choices rather than from four. When answering the questions that require an explanation, express your thoughts clearly so that the grader will know whether you understand the concept being tested. Remember to write legibly because the grader may not be as able to interpret your handwriting as well as your teacher.

The solutions provided for these questions are not the only possible solutions because there is often more than one way to solve a problem.

THE SAMPLE TEST

SESSION I—PART I

1. In which order should you perform the operations to simplify this expression?

$$6 + 8(11 - 2) \div 3$$

   A. $+, \times, -, \div$
   B. $\times, -, +, \div$
   C. $-, \times, \div, +$
   D. $-, \times, +, \div$

2. What value of k will make this a true statement?

$$3^k = 243$$

   F. 3
   G. 2
   H. 4
   J. 5

3. What is the prime factorization of 108?

   A. $3^3 \times 2^2$
   B. $6 \times 3^2 \times 2$
   C. $3^2 \times 2^3$
   D. $2^3 \times 3^3$

4. Del and Tony are mowing their neighbor's lawn. Del mowed 20% and Tony mowed $\frac{3}{5}$ of the lawn. What percent is left to mow?

   F. 20%
   G. 30%
   H. 25%
   J. 35%

5. The gas tank of a car was filled before a trip. After the trip, the gas gauge indicator pointed exactly between $\frac{1}{2}$ and $\frac{3}{4}$. What fraction of the gas was used on this trip?

A. $\frac{3}{8}$

B. $\frac{5}{8}$

C. $\frac{3}{4}$

D. $\frac{1}{3}$

6. The drawing below shows an isosceles triangle placed on a square.

The lengths of the congruent sides of the triangle are 9. The perimeter of this figure is 60. What is the *area* of the square?

F.　36
G.　56
H.　196
J.　360

7. Line segments PT, MW, and RZ intersect, forming six angles. Based on this diagram, which of the following statements is true?

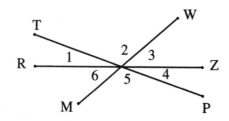

    A. Angles 3 and 4 are supplementary.
    B. Angle 1 is congruent to angle 4.
    C. Angle 2 and angle 3 are complementary.
    D. Segments RZ and PT are congruent.

8. Use your ruler for this question. An odd-shaped lot needs fencing.

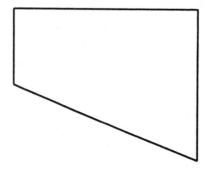

    Based on the scale of 1 centimeter = 3 meters, how much fencing is needed to enclosed the lot?

    F. 16.5 meters
    G. 49.5 meters
    H. 150 meters
    J. 330 meters

9. The school parking lot was $\frac{3}{8}$ empty. What percentage of the lot was full?

    A. 58%

    B. 75%

    C. 62.5%

    D. 37.5%

10. Sally and Kira are selling Girl Scout cookies. Sally says to Kira, "If I sell this box and 3 more, I will have a total of 12 sold." Kira replies, "If I sell 4 more boxes, I will have sold 16 boxes." Which of the following statements is correct?

    F. Kira and Sally sold 15 boxes.

    G. Sally sold 8 boxes.

    H. Sally sold 3 more boxes than Kira.

    J. Kira sold 7 boxes.

11. Which value of x will make this statement true?

$$6(x - 7) - 5 = 19$$

    A. 9

    B. 11

    C. 13

    D. 15

12. The new hotel has a fountain in the front. The fountain is square.

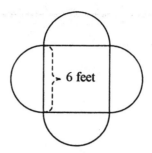

On each side of the square is a semicircular flowerbed. What is the perimeter of the flowerbeds to the nearest foot?

    F.   24 feet

    G.   20 feet

    H.   38 feet

    J.   54 feet

13. The postage rates for mailing a box are $2.70 for the first 3 pounds and $.20 for each ounce thereafter. What would it cost to mail a 4.5-pound box?

    A. $3.85

    B. $5.10

    C. $6.70

    D. $7.50

14. Based on your knowledge of geometry, which of the following statements is false?

    F.   The sum of the acute angles in a right triangle is complementary.

    G.   All squares are rectangles.

    H.   The diagonals of a parallelogram are not always congruent.

    J.   The area of a circle is a little more than three times its diameter.

15. Which of the following inequalities is graphed correctly?

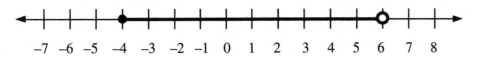

-7 -6 -5 -4 -3 -2 -1 0 1 2 3 4 5 6 7 8

A. $-4 < x \geq 6$
B. $-4 \leq x \leq 6$
C. $-4 < x \leq 6$
D. $-4 \leq x < 6$

16. The hypotenuse of a right triangle is $\sqrt{29}$. If one leg measures 5, what is the length of the other leg?

F. 4
G. 3
H. 5
J. 2

17. The spinner is spun once. What is the probability that you will *not* get a vowel?

A. $\frac{3}{8}$

B. 25%

C. $\frac{5}{8}$

D. 62%

18. Which of the following is an irrational number?

    F. $\sqrt{81}$

    G. $\sqrt{\dfrac{1}{4}}$

    H. $\dfrac{\pi}{3}$

    J. $5.\overline{3}$

19. The video rental store sells previously viewed movies for $10 each. The sign over the display reads:

Buy	Save
2	$4
3	$8
4	$12
5	$16

    At this rate, how many videos can you buy with $60?

    A. 6
    B. 8
    C. 9
    D. 10

20. After a reflection over the *x*-axis and then a reflection over the *y*-axis, a trapezoid's image appears in the fourth quadrant. What are the other two vertices of the original trapezoid?

   F.   (−5, −2) and (−3, −2)
   G.   (−9, 7) and (−1, 7)
   H.   (−5, 4) and (−1, 4)
   J.   (−5, 4) and (−3, 4)

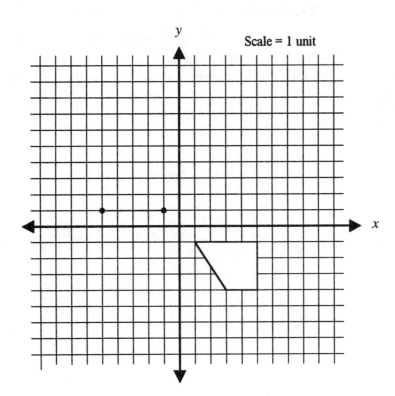

Scale = 1 unit

21. The average of five numbers is 65. What number can you add to this list of five numbers so the average of all six numbers is 70?

   A. 65
   B. 90
   C. 70
   D. 95

22. In the dry winter months, the classroom aquarium loses $\frac{1}{4}$ inch of water per week due to evaporation. Let f represent the height of the water in a full aquarium and let l represent the amount of water after a loss. Which equation best describes s, the amount of water in the aquarium after six weeks with only 1 inch of water added during that same period?

   F. $f - s = \frac{1}{2}$

   G. $f - \frac{6}{4} = s$

   H. $\frac{3}{2} + 1 = s$

   J. $f + 1 = s$

Answer questions 23 and 24 using the following table.

Name	Number of Assignments
Ari	17
Bart	16
Casey	19
Deb	20
Derek	17
Erin	18
Georgia	17
Manny	15
Matt	16
Mike	19
Tara	20

23. The table shows the number of perfect homework out of 20 assignments completed in Mr. Peterson's math class.

What is the *median* number of perfect assignments?

A. 16
B. 17
C. 17.6
D. 18

24. Jillian was left off the list in error. What is the minimum number of perfect assignments she needs so that the *mean* of the group is 17.5?

F. 15
G. 16
H. 17.5
J. 19

25. Joey has a part-time job on Saturdays at the grocery store folding advertising fliers. He gets $3.50 per hour plus $.02 per flier. On the average, Joey can fold two fliers per minute. How many hours did he work last Saturday if his check was $38.35?

   A. 11 hours
   B. 10 hours
   C. 7 hours
   D. 6.5 hours

26. When Helen babysat her neighbor's daughter, Chelsea, she brought cookies. Chelsea ate $\frac{1}{4}$ of the cookies. Of what Chelsea left, Helen ate $\frac{2}{3}$. Six cookies were left on the plate. How many cookies did Helen bring?

   F. 12
   G. 18
   H. 24
   J. 36

27. The scale listed on the box of a model ship is 1 inch:8 feet. Using a proportion, how can you find the actual length of the ship if the length of the model is 5 feet?

   A. $\frac{1}{8} = \frac{x}{60}$

   B. $\frac{1}{8} = \frac{60}{x}$

   C. $\frac{x}{8} = \frac{5}{60}$

   D. $\frac{1}{8} = \frac{5}{x}$

SESSION I—PART II

28. Which of the following numbers can replace x in the inequality?

$$\frac{7}{18} < x < 0.68$$

Circle all the numbers that will make the inequality true.

$$\frac{2}{3} \qquad 0.72 \qquad 0.6 \qquad \frac{5}{7}$$

Explain why each number you circled can replace x.

29. This square is divided into 12 regions. The small regions are the result of dividing a larger region in half. Find the fractions represented by the letters relative to the entire square. Explain your answers.

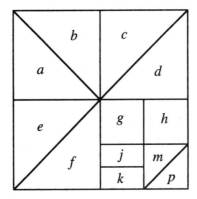

A. $c =$

B. $p =$

C. What is the ratio of the area of m to the area of b? Explain how you arrived at your answer.

30. The corner ice cream parlor has the following menu posted.

Cone	Ice Cream	Toppings
Regular	Chocolate	Chocolate jimmies
Sugar	Vanilla	Chopped nuts
Waffle	Strawberry	Bubble gum chips
	Cookie dough	Confetti jimmies
	Black raspberry	

    A. How many different cones can be made? Show your work.

    B. What is the probability of getting cookie dough ice cream without chocolate jimmies in any kind of cone? Show your work.

31. An 80-foot flag pole casts a 30-foot shadow. Find the height of the tree if its shadow is 9 feet. Show your work.

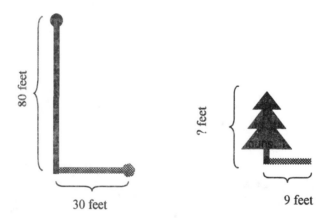

32. Before you are nine identical looking gems. Eight gems weigh exactly the same. Explain how you can determine which gem is the heavier one using a balance scale twice only.

33. There are 73 students in the eighth grade class at Garfield Middle School. Eleven do not participate in after-school activities. Of the remaining students, 35 participate in volleyball (V) and 39 are members of the Math-Science-Computer Club (MSC). How many students participate in both volleyball and the club? Fill in the correct numbers in the Venn diagram and explain your answer.

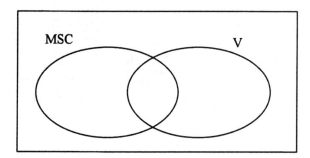

SESSION II

The eighth grade class was in charge of School Spirit Week's Day at the Fair. The money raised would help purchase new library books. In addition to planning the fair, they were in charge of the snack booth and the fishing booth.

Answer questions 34, 35, and 36 about Day at the Fair.

34. In the fishing booth, Al was under a skirted table and attached a toy to the end of the fishing line. He purchased a bag of small toys for $7.00. Each fishing attempt cost the student $.25. All the toys in the bag cost $.10 each. How many students were needed to play before the booth started making a profit? Explain your answer.

35. Jamie was in charge of the snack booth, which made a profit of $120.30. She sold 126 orange drinks at $.75 each and sold cookies for $.60 each.

   A. Write an equation to find the number of cookies sold.
   B. Solve the equation and show your work.

36. The report for Day at the Fair follows.

Booth	Cost	Profit	Net Profit
Pie booth	$10.00	$107.50	$97.50
Fishing booth	$7.00		
Snack booth	$9.85	$120.30	$110.45
Bean bag toss		$42.70	
Bear booth	$25.00	$63.40	
Total	$57.35	$362.15	

   A. Angela didn't finish calculating the results before the committee meeting. Complete the table.
   B. What is the amount that can be used to purchase the library books if shipping charges are 15% of the order? Show your work.

37. Two of the vertices of an isosceles triangle are (–5, –3) and (–1, 6). Find the coordinates of the third vertex so that the area of this triangle is 36 square units. Graph the triangle on the grid, and explain your answer.

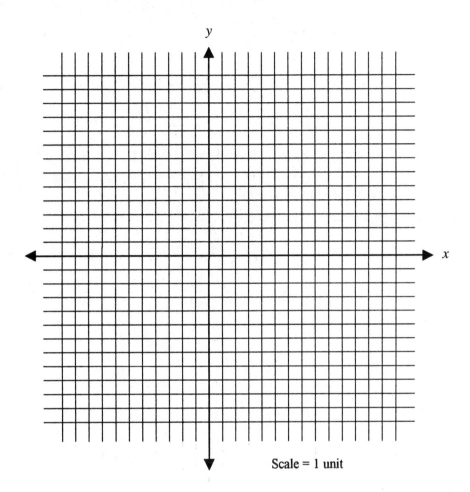

Scale = 1 unit

Answer questions 38, 39, and 40 about the aluminum cans.

A company makes aluminum cans that are cylindrical. The height is 5 cm and the circular base has a radius of 2 cm. The company is discussing changing the shape of the cans so that they have a square base and a height of 3 cm.

38. Find the volume of the cylindrical can to the nearest centimeter. Show all your work.

39. Find the length of the square base to the nearest centimeter of the new can design that has a height of 3 cm. This new can should have approximately the same volume as the cylindrical can. Show your work.

40. Which can uses more aluminum, the cylinder or the new design? Explain how you arrived at your answer.

41. A. Find the length of the hypotenuse in right triangle ABC. Show your work.
   B. Find the tangent of angle B. Show your work.

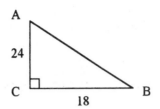

42. Scale A and scale B are in perfect balance. How many ✦s will be needed to balance scale C? Explain your answer.

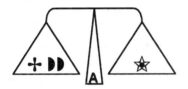

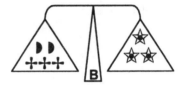

43. A rectangle is inscribed in a quarter of a circle. The radius of the circle measures 13 units. Line segments AB = 12 units and ED = 1 unit. Find the length of $\overline{AD}$. Explain your answer.

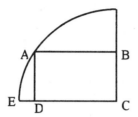

44. The quarterly profits of a store called *The Craft Corner* are shown in the following table. At this rate, estimate how many more quarters it will take for the profits to reach $10,000. Explain your answer.

Dates	Income
January 1 to March 31	$600
April 1 to June 30	$580
July 1 to September 30	$530
October 1 to December 31	$680

45. A list containing the average high temperatures for Buffalo as well as the record high temperatures for each month follows.

A. Draw a double line graph that compares these temperatures. Be sure to title the graph, label the axes, use appropriate scales, graph correctly, and create a key.

B. After completing your graph, write a sentence or two about your observation.

Month	Average High	Record High
January	30°F	72°F
February	33°F	69°F
March	43°F	81°F
April	56°F	89°F
May	68°F	91°F
June	76°F	96°F
July	81°F	103°F
August	78°F	100°F
September	71°F	96°F
October	60°F	88°F
November	47°F	80°F
December	35°F	73°F

SAMPLE TEST SOLUTIONS

SESSION I—PART I

1. **C.** The order of operations (PEMDAS) states that parentheses are solved first. This means that choices A and B should not be considered.

The expression:	$6 + 8(11 - 2) \div 3$
After parentheses:	$6 + 8(9) \div 3$
After multiplication:	$6 + 72 \div 3$
After division:	$6 + 24$
After addition:	30

2. **J.** Choices F and G are too small to make sense. You know $3^4 = 81$. That leaves 5 as the correct exponent.

3. **A.** Make a factor tree.

$$108 = 2^2 \times 3^3$$

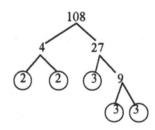

4. **F.** Change $\frac{3}{5}$ to 60%. Add 60% to 20%. 100% − 80% leaves 20% left to mow.

5. **A.** Find the fraction between $\frac{1}{2}$ and $\frac{3}{4}$ by writing both fractions with a common denominator of 8.

$$\frac{1}{2} < x < \frac{3}{4} \quad \text{or} \quad \frac{4}{8} < x < \frac{6}{8}$$

$$x = \frac{5}{8}$$

The indicator shows the amount of gas in the tank now. This means that $\frac{3}{8}$ of the tank was used on the trip.

6. **H.** Two sides of the pentagon have a length of 9 units and three sides have the same length. Write an equation for the perimeter of the pentagon where x is the length of one of the three congruent sides of the pentagon.

$$2(9) + 3x = 60. \text{ (This is the same as } 9 + 9 + x + x + x = 60.)$$

$$18 + 3x = 60$$

$$3x = 42$$

$$x = 14$$

The area of the square is 14^2 or 196 square units.

7. **B.** Look at each choice. Do not assume that there are right angles in the diagram unless that fact is specifically stated in the problem.
   A is false because the sum of angles 3 and 4 is definitely less than 180°. C is false because the angle formed by 2 and 3 is greater than a right angle. You cannot be sure about D based on the information given. Angles 1 and 4 are vertical angles, which make then congruent.

8. **G.** The perimeter of the trapezoid is 16.5 cm. If each centimeter = 3 meters, the perimeter of the lot is 16.5 × 3 or 49.5 meters.

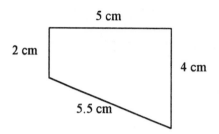

9. **C.** The parking lot was $\frac{3}{8}$ empty or $\frac{5}{8}$ full. Change $\frac{5}{8}$ to a decimal (0.625) and then to a percent. So the lot is 62.5% full.

10. **G.** From Kira's statement you know she sold 12 boxes because 12 + 4 = 16. Sally sold 8 boxes because "this box" + 3 + 8 = 12. This is stated in choice **G.**

11. **B.** Solve this equation:

$$6(x - 7) - 5 = 19$$
$$6(x - 7) = 24$$
$$(x - 7) = 4$$
$$x = 11$$

12. **H.** The perimeter of the flowerbeds is found by finding the circumference of a circle with a diameter of 6 feet, then doubling your answer. Four semicircles is equivalent to two full circles.

$C = \pi d$ (the formula to find the circumference of a circle)

$C \approx (3.14)6 \approx 18.8$, which is the circumference of one circle.

Use 3.14 as an approximation for pi.

So, 18.8 × 2 = 37.6 or 38 feet is the perimeter of the flowerbeds.

13. **D.** After \$2.70 for the first 3 pounds are charged, there are 1.5 pounds or 24 ounces yet to pay (16 ounces = 1 pound). The cost would be \$2.70 + 24(\$.20) = \$2.70 + \$4.80 = \$7.50.

14. **J.** You are looking for the *false* statement. Statement J can be made true by changing *area* to *circumference* or else by changing *is a little more than three times its diameter* to π *times the radius squared.*

15. **D.** Look at the graph and see that −4 is included (the darkened circle) but 6 is not (the open circle). **D** has the correct notation.

16. **J.** Use the Pythagorean Theorem: $a^2 + b^2 = c^2$.

$$5^2 + x^2 = (\sqrt{29})^2$$

$$25 + x^2 = 29$$

$$x^2 = 4$$

$$x = 2$$

17. **C.** The probability of not getting a vowel is the same as the probability of getting a consonant, which is 5 out of 8 or $\frac{5}{8}$.

18. **H.** $\sqrt{81} = 9$ and $\sqrt{\frac{1}{4}} = \frac{\sqrt{1}}{\sqrt{4}} = \frac{1}{2}$. That disqualifies choices F and G. In J, $5.\overline{3} = 5.33333...$, which is a rational number. The only irrational number is **H.** The symbol π should have been a hint!

19. C. Add another column to list the amounts you would spend. For each extra video you buy, you spend an extra $6. You can buy at most 9 videos.

Buy	Save	Spend
2	$4	$20 − 4 = $16
3	$8	$30 − 8 = $22
4	$12	$40 − 12 = $28
5	$16	$50 − 16 = $34
6		$34 + 6 = $40
7		$40 + 6 = $46
8		$46 + 6 = $52
9		$52 + 6 = $58
10		$58 + 6 = $64

20. J.

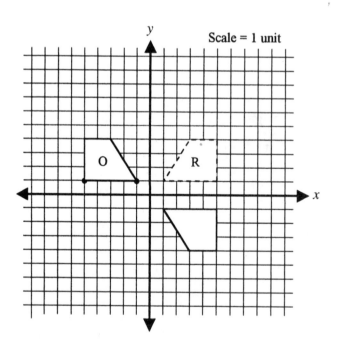

Work backward. Reflect the trapezoid over the *x*-axis. This is shown by the trapezoid labeled R. Then reflect R over the *y*-axis. Draw in the original trapezoid to find the coordinates of the other two vertices of (−5, 4) and (−3, 4).

21. **D.** The sum of those five numbers has to be 5×65 or 325 to get an average of 65. The sum of 6 numbers with an average of 70 is 6×70 or 420. You must add 95 to 325 to get 420.

22. **F.** Over six weeks, the aquarium lost 1.5 inches of water. The total loss was 0.5 inch since 1 inch of water was added. Starting with a full tank, f, subtract 0.5 inch of water to get s.

$$f - \frac{1}{2} = s \quad \text{or} \quad f - s = \frac{1}{2}$$

23. **B.** To find the median, arrange the numbers in order. The median, or the sixth value, is 17.

24. **G.** To get an average of 12 numbers to be 17.5, the sum of those numbers has to be 210 ($17.5 \times 12 = 210$). The sum of the 11 numbers in the table is 194. You must add 16 to 194 to get 210.

25. **D.** If Joey folds two fliers per minute, he folds 120 per hour. At an extra $.02 cents per flier, he adds $2.40 to his hourly rate of $3.50.

$$\$38.35 \div \$5.90 \text{ per hour} = 6.5 \text{ hours}$$

26. **H.** Guess and check would be one way to solve this problem. If Helen brought 12 cookies, it means Chelsea ate 3 and left 9 on the plate for Helen. Helen ate 6 and left 3 on the plate. So, 12 is not a choice. If Helen brought 18, then Chelsea ate $4\frac{1}{2}$ cookies and left $13\frac{1}{2}$ for Helen. Helen ate 9 and left $4\frac{1}{2}$ cookies on the plate. So, 18 is not a choice. If Helen brought 19, Chelsea ate 9 and left 27 and Helen ate 9 and left 18. So, 36 is not a choice.

Working backward is another way to solve it. If Helen ate two thirds of the cookies, then one third or 6 cookies were left. This means that there were 18 cookies left on the plate after Chelsea ate some. If Chelsea ate one quarter, then three quarters or 18 cookies were left. This means that Helen brought 24 cookies.

27. **B.** Keep the numbers of the model size in the numerator and the full-sized ship in the denominator. The model size is 1 inch to 8 feet. You know the model is 5 feet long, but you must convert to inches.

$$\frac{1 \text{ inch}}{8 \text{ feet}} = \frac{60 \text{ inches}}{x \text{ feet}}$$

$$x = 8(60)$$

$$x = 480 \text{ feet}$$

SESSION I—PART II

28. Change all the fractions to decimals. Now the problem becomes

$$0.3888... \; < \; x \; < \; 0.68$$

$$\frac{2}{3} = 0.666... \qquad \frac{5}{7} = 0.714...$$

The numbers that can replace x are $\frac{2}{3}$ and 0.6. When the numbers are written in the same form, it is easy to see that these two numbers lie between 0.38888... and 0.68.

29. **A.** The value of c is $\frac{1}{8}$ because the large square is cut into fourths and c is one half of one quarter, or $\frac{1}{2} \times \frac{1}{4} = \frac{1}{8}$.

B. The value of p is $\frac{1}{32}$ since g is one fourth of a fourth, or $\frac{1}{16}$, and p is $\frac{1}{2}$ of $\frac{1}{16}$, or $\frac{1}{32}$.

C. Since b and c are congruent and m and p are congruent, the ratio of $m{:}b$ is $\frac{1}{32}{:}\frac{1}{8}$, or 1:4.

30. **A.** There are $3 \times 5 \times 4 = 60$ different cones.

B. There are $3 \times 1 \times 3 = 9$ different cones. The probability of getting these cones is $\frac{9}{60} = \frac{3}{20}$.

31. Use a proportion to solve for the height of the tree by comparing the shadows of the objects to their heights. Notice that the "tree" is in both numerators and the "shadow" is in both denominators.

$$\frac{\text{Shadow of tree}}{\text{Shadow of pole}} = \frac{\text{Height of tree}}{\text{Height of pole}}$$

$$\frac{9}{30} = \frac{x}{80}$$

$$30x = 720$$

$$x = 24 \text{ feet}$$

You can also set up the proportion with the ratio of height:shadow.

$$\frac{\text{Height of pole}}{\text{Shadow of pole}} = \frac{\text{Height of tree}}{\text{Shadow of tree}}$$

$$\frac{80}{30} = \frac{x}{9}$$

$$30x = 720$$

$$x = 24 \text{ feet}$$

Either way, the height of the tree is 24 feet.

32. Divide the nine gems into three piles of three gems. Put one pile on each side of the balance scale, setting aside the third pile. If the scale is balanced, you know that the heavier gem is in the third pile. If the scale is not balanced, the heavier gem is on the side of the scale that is lower. Now you know which pile of three gems contains the heavier one. Disregard the other six gems. For the second weighing, place one gem on each side of the scale. If the scale is balanced, the heavier gem is not on the scale. If the scale is not balanced, the heavier gem is on the scale. Yes, out of nine gems you can find the heavier one in two weighings.

33. Of the 73 students, 73 – 11 or 62 students participate in activities. If 39 are in MSC and 35 in V, 39 + 35 = 74. That means 74 – 62 or 12 students are in both clubs. So 12 goes in the overlapping region. Since 39 are in MSC and 12 are accounted for, then 27 participate in MSC only. With 35 in V and 12 accounted for, then 23 are in V only. The 11 nonparticipating students are outside the circles. To check your answer, add the numbers and make sure the total is 73 (27 + 12 + 23 + 11 = 73).

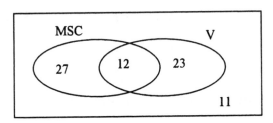

SESSION II—PART II

34. Each student paid 25 cents for a 10-cent toy for one try at the fishing booth. To start making a profit, $7.00 ÷ $.25 = 28. So, 28 students had to play to cover the cost of the toys.

35. A. Let x represent the number of cookies sold.

126 drink at $.75 + x cookies at $.60 = $120.30 booth income

$$126(.75) + x(.60) = 120.30$$

This portion of the question wants only the equation. Solving this equation is done in **B.**

B.

$$126(.75) + .6x = 120.3$$

$$94.5 + .6x = 120.3$$

$$945 + 6x = 1203$$

$$6x = 258$$

$$x = 43$$

There were 43 cookies sold.

36. A. Find the cost of the bean bag toss by adding the costs from the other booths, then subtracting from $57.35. The same procedure can be done to find the profit from the fishing booth.

Booth	Cost	Profit	Net Profit
Pie booth	$10.00	$107.50	$97.50
Fishing booth	$7.00	$28.25	$21.25
Snack booth	$9.85	$120.30	$110.45
Bean bag toss	$5.50	$42.70	$37.20
Bear booth	$25.00	$63.40	$38.40
Total	$57.35	$362.15	$304.80

B. The proceeds from Fair Day can be used to purchase new library books. If 15% of this total will be used for shipping, then 85% of this total is the amount that can be spent on books.

$$304.80 (.85) = \$259.08 \text{ is the limit}$$

37. Here are two possible triangles. The gray triangle's third vertex is (3, –3) and the white triangle's vertex is (–9, 6).

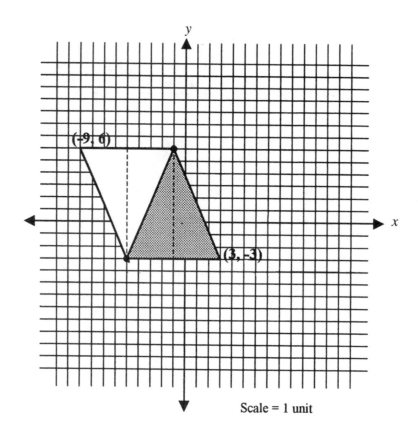

Scale = 1 unit

Use the formula for the area of a triangle. The height of the triangles is denoted by the dotted line, which is nine units. Find the base.

$$A = \frac{bh}{2}$$

$$36 = \frac{9b}{2}$$

$$72 = 9b$$

$$8 = b$$

When you have the length of the base, draw in the rest of the triangle. Find the coordinates of the third vertex.

38.

$$V = \pi r^2 h$$

$$V = \pi(2^2)(5)$$

$$V \approx 3.14(4)(5) \approx 62.8$$

The volume of the cylindrical can is about 63 cm³.

39. The volume of the new can is *Length* × *Width* × *Height*. The length and width will be the same because of the square base. The volume of both cans should be approximately the same.

$$V = lwh$$

$$V = l^2 h$$

$$63 = l^2(3)$$

$$21 = l^2$$

$$l \approx 4.6$$

So, the base of the new can is a 5-inch square.

40. Find the surface area of both cans, then compare.

The surface area of the cylindrical can is the sum of the area of the top and bottom circles and the rectangular region under the label.

$$A = \pi r^2$$

$$A = \pi(2^2)$$

$$A \approx 12.6 \text{ cm}^2$$

The top of the can has an area of approximately 12.6 cm². The area under the label is a rectangle where the length is the circumference of the can's top and the height is 5 cm.

$$A = \pi d h$$

$$A = \pi(4)(5)$$

$$A \approx 62.8 \text{ cm}^2$$

The total surface area of the cylindrical can is 2(12.6) + 62.8 or 88 cm².

The surface area of the new design is the sum of the areas of each of the six faces. Top and bottom dimensions: length of 5 cm and width of 5 cm (square base). Four sides have the same dimensions: length of 5 cm and height of 3 cm.

$$2(5)(5) + 4(5)(3) = 50 + 60 = 110 \text{ cm}^2$$

The new can will use 22 cm² more aluminum than the old can.

41. A. Use the Pythagorean Theorem to find the hypotenuse.

$$c^2 = a^2 + b^2$$

$$c^2 = 24^2 + 18^2$$

$$c^2 = 576 + 324 = 900$$

$$c = 30 \text{ units}$$

B. Use the trigonometric function ratio to find the tangent.

$$\tan = \frac{\text{Opposite}}{\text{Adjacent}} = \frac{24}{18} = \frac{4}{3} = 1.\overline{3}$$

42. The question mark should be replaced by four ✤ symbols. The total number of stars from the right side of scales A and B appear on the left side of scale C. If you combine the symbols from the left sides of scales A and B, you will see that four ✤ symbols are required.

43. $\overline{AB} = \overline{DC} = 12$ because ABCD is a rectangle and the opposite sides are congruent. $\overline{AC} = 13$ because it is a radius of the circle. Triangle ADC is a right triangle. Use the Pythagorean Theorem to solve for $\overline{AD}$.

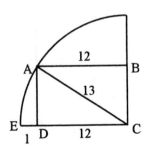

$$a^2 + b^2 = c^2$$

$$a^2 + 12^2 = 13^2$$

$$a^2 = 169 - 144$$

$$a^2 = 25$$

So $\overline{AD} = 5$. You may also remember that 5-12-13 is a Pythagorean triple.

44. *The Craft Corner* has an income of approximately $600 per quarter, found by averaging the four quarters. The annual income can be estimated at $2,400. At the end of four years (16 quarters), the estimated income will be $9,600. The income of $10,000 will be reached by the end of the 17th quarter. Since the question asks "how many *more* quarters will it take," the answer is 13 more. Do not count the original four quarters.

45. A.

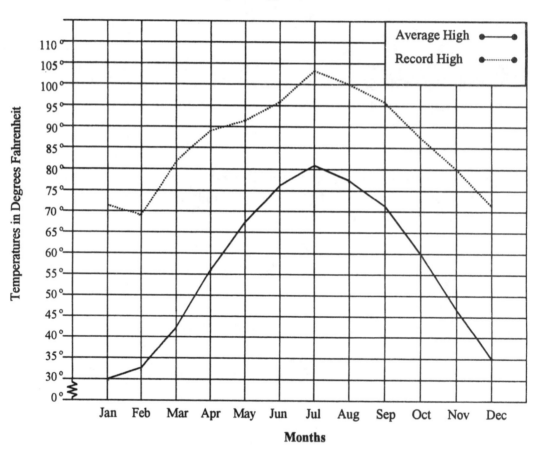

B. The space between the lines is greater in the winter than in the summer. You can say that the difference between the temperatures in summer (about 25°) is less than it is in winter (about 40°).

NOTES

NOTES

NOTES

NOTES

NOTES

NOTES

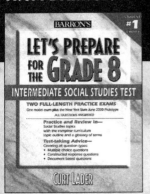
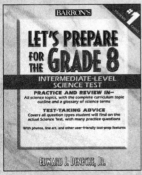
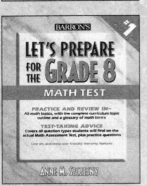

SHAKESPEARE MADE EASY

Macbeth

MODERN ENGLISH VERSION
SIDE-BY-SIDE WITH FULL ORIGINAL TEXT

BARRON'S

At last! Shakespeare in Language everyone can understand... SHAKESPEARE MADE EASY Series

Scene 7

Macbeth's castle. Enter a **sewer** *directing divers servants. Then enter* **Macbeth**.

Macbeth If it were done, when 'tis done, then 'twere well
    It were done quickly: if th' assassination
    Could trammel up the consequence, and catch,
    With his surcease, success; that but this blow
5   Might be the be-all and the end-all here,
    But here, upon this bank and shoal of time,
    We'd jump the life to come. But in these cases
    We still have judgement here: that we but teach
    Blood instructions, which being taught return
10 To plague th'inventor: this even-handed justice
    Commends th'ingredience of our poisoned chalice
    To our own lips. He's here in double trust:
    First, as I am his kinsman and his subject,
    Strong both against the deed: then, as his host,
15 Who should against his murderer shut the door,
    Not bear the knife myself. Besides, this Duncan
    Hath borne his faculties so meek, hath been
    So clear in his great office, that his virtues
    Will plead like angels, trumpet-tounged, against
20 The deep damnation of his taking-off;
    And pity, like a naked new-born babe,
    Striding the blast, or Heaven's cherubin, horsed
    Upon the sightless couriers of the air,
    Shall blow the horrid deed in every eye,
25 That tears shall drown the wind. I have no spur
    To prick the sides of my intent, but only
    Vaulting ambition, which o'erleaps itself,
    And falls on th'other –

Scene 7

A room in **Macbeth's** *castle. A* **Butler** *and several* **Waiters** *cross, carrying dishes of food. Then* **Macbeth** *enters. He is thinking about the proposed murder of* **King Duncan**.

Macbeth If we could get away with the deed after it's done, then the quicker it were done, the better. If the murder had no consequences, and his death ensured success...If, when I strike the blow, that would be the end of it – here, right here, on this side of eternity – we'd willingly chance the life to come. But usually, we get what's coming to us here on earth. We teach the art of bloodshed, then become the victims of our own lessons. This evenhanded justice makes us swallow our own poison. [*Pause*] Duncan is here on double trust: first, because I'm his kinsman and his subject (both good arguments against the deed); then, because I'm his host, who should protect him from his murderer—not bear the knife. Besides, this Duncan has used his power so gently, he's been so incorruptible his great office, that his virtues will plead like angels, their tongues trumpeting the damnable horror of his murder. And pity, like a naked newborn babe or Heaven's avenging angels riding the winds, will cry the deed to everyone so that tears will blind the eye. I've nothing to spur me on but high-leaping ambition, which can often bring about one's downfall.

A simplified modern translation appears side-by-side with the original Elizabethan text...plus there's helpful background material, study questions, and other aids to better grades.

   Yes, up-to-date language now makes it easier to score well on tests *and* enjoy the ageless beauty of the master's works.

Shakespeare is Made Easy for these titles:

Hamlet, $6.95
Henry IV, Part One, $6.95
Julius Caesar, $6.95
King Lear, $6.95
Macbeth, $6.95
The Merchant of Venice, $6.95

A Midsummer's Night's
   Dream, $6.95
Romeo & Juliet, $6.95
The Tempest, $6.95
Twelfth Night, $6.95

BARRON'S

Barron's Educational Series, Inc.
250 Wireless Boulevard
Hauppauge, New York 11788

(#16) R 1/04

Really. This isn't going to hurt at all . . .

Barron's *Painless* titles are perfect ways to show kids in middle school that learning really doesn't hurt. They'll even discover that grammar, algebra, and other subjects that many of them consider boring can become fascinating— and yes, even fun! The trick is in the presenta-tion: clear instruction, taking details one step at a time, adding a light and humorous touch, and sprinkling in some brain-tickler puzzles that are both challenging and entertaining to solve.

Each book: Paperback, approx. 224 pp., $8.95–$10.95 Canada $11.95–$15.95

Painless Algebra
Lynette Long, Ph.D.
ISBN 0-7641-0676-7

Painless American History
Curt Lader
ISBN 0-7641-0620-1

Painless Fractions
Alyece Cummings
ISBN 0-7641-0445-4

Painless Geometry
Lynette Long, Ph.D.
ISBN 0-7641-1773-4

Painless Grammar
Rebecca S. Elliott, Ph.D.
ISBN 0-8120-9781-5

Painless Math Word Problems
Marcie Abramson, B.S., Ed.M.
ISBN 0-7641-1533-2

Painless Poetry
Mary Elizabeth
ISBN 0-7641-1814-2

Painless Research Projects
*Rebecca S. Elliott, Ph.D.,
and James Elliott, M.A.*
ISBN 0-7641-0297-4

Painless Science Projects
Faith Hickman Brynie, Ph.D.
ISBN 0-7641-0595-7

Painless Speaking
Mary Elizabeth
ISBN 0-7642-2147-2

Painless Spelling
Mary Elizabeth
ISBN 0-7641-0567-1

Painless Writing
Jeffrey Strausser
ISBN 0-7641-1810-2